The HAYMAKERS

The HAYMAKERS

A Chronicle of Five Farm Families

STEVEN R. HOFFBECK

www.mnhs.org/mhspress

Manufactured in the United States of America.

10 9 8 7 6 5 4 3 2 1

International Standard Book Number
0-87351-394-0 (cloth)
0-87351-395-9 (paper)

∞ The paper used in this publication meets the minimum requirements of the American National Standard for Information Sciences—Permanence for Printed Library Materials, ANSI Z39.48-1984.

Library of Congress Cataloging-in-Publication Data

Hoffbeck, Steven R.
The haymakers : a chronicle of five farm families / Steven R. Hoffbeck.
    p. cm.
Includes bibliographical references (p.).
ISBN 0-87351-394-0 (cloth : alk. paper)
ISBN 0-87351-395-9 (pbk. : alk. paper)
    1. Hay—Harvesting—Minnesota—History.
    2. Farmers—Minnesota—History.
    I. Title.

SB198 .H576 2000
633.2′085′09776—dc21
00-033930

Title pages: *Family mowing and raking hay in Morrison County, about 1900.*

This book is for

Dianne,

for my children, Leah, Katie, Mary, and John W.

for my mother, Alvina Elsie Emilie (Engel) Hoffbeck

and in memory of
Raymond Peter Hoffbeck and
Larry William Hoffbeck

The HAYMAKERS

What is a farm but a chapter in the bible almost? Pull out the weeds, water the plants; blight, rain, insects, sun,—it is mere holy emblem from its first process to the last.

RALPH WALDO EMERSON
Journals and Miscellaneous Notebooks, 1836

The HAYMAKERS

Mowing

There was never a sound beside the wood but one,
And that was my long scythe whispering to the ground.
What was it it whispered? I knew not well myself;
Perhaps it was something about the heat of the sun,
Something, perhaps, about the lack of sound—
And that was why it whispered and did not speak.
It was no dream of the gift of idle hours,
Or easy gold at the hand of fay or elf:
Anything more than the truth would have seemed too weak
To the earnest love that laid the swale in rows,
Not without feeble-pointed spikes of flowers
(Pale orchises), and scared a bright green snake.
The fact is the sweetest dream that labor knows.
My long scythe whispered and left the hay to make.

ROBERT FROST

Prologue

We always kept track of close calls on the farm. It was Dad's way of teaching us to be careful. Once, when I was still too young to remember, the combine caught his shirttail in a moving belt. Because he was strong and the shirt was cotton, it ripped and he wasn't hurt, but what if it had been thick denim? Mom took a picture as a reminder of how close he had come, a permanent image in the family photo album from that time on. The picture appeared next to snapshots of birthdays and anniversaries: Dad, younger than any of us ever knew him, bare-chested but for shreds of his ribboned workshirt dangling down to his waist. He stands posed beside the combine: one hand in his pocket, hat tipped slightly to the side—positioned next to the exposed belt and sprocket that nearly pulled him in. His face is clenched in a grimace, but his chest is strong and unharmed.

The lesson was clear: watch out for that machine; watch your step, your hands, and your feet around the combine, the hammer-mill grinder, the mowing machine, the corn picker and the corn sheller, the silage blower and the haybale elevator. Be careful that you don't get your hand pinched hooking up the hayrack to the tractor hitch. Move with caution, every day, on every inch of the Hoffbeck farm.

We were taught to be especially careful around moving parts. The worst I faced was the power-takeoff (PTO) shaft we ran from the tractor to the hammer-mill to grind corn and oats into prime feed for the cows. It was just a bare PTO shaft, no shields, parked by the feed room. I took care to climb up the tractor, stepping past and over the shaft to get to the seat, where I cranked the throttle up to grinding speed. I always looked and stepped wisely to avoid it on the way down, careful to the point of hugging, literally, the tractor's fender. Everyone feared the PTO, for everyone had heard of someone who had been caught. I didn't know *how* it killed someone, but I believed it was awful. These are the fears of farmkids.

In late November 1968, the worst of those fears was realized. It had already been a tough fall for Dad; the autumn skies brought too much rain and the soybean fields were too wet to harvest in September. The rains continued, and the beanfields were still too wet even through all of October. Finally in November, after the ground had frozen, Dad could work with the combine in the field, but it was so cold that he had to wear more warm clothes than usual. Whether from the fatigue of the drawn-out harvest or from the frustration of working with an aging combine-harvester, my dad made a mistake on the afternoon of the nineteenth, and somehow he got his bulky clothing caught in the PTO shaft.

My mom prepared an afternoon lunch for us kids to eat when we got home from school, wondering why Dad hadn't walked home for coffee at four o'clock like he usually did. She finally went through the grove on foot to see what was keeping him. His body had been wrapped up tight in the twisting of his clothes by the PTO shaft, and he had suffocated. I was at basketball practice, in town, when my mom ran home to call an ambulance. They came and cut him from the mechanical grip of the combine, but he was dead on arrival at the Redwood Falls hospital. When I got home, my sister Ane Marie was the only one there waiting for me, tears on her face.

"Daddy died." That was all she said.

In the silence of the morning following the accident, and with none of us going to school that day, I went out to the north of the grove to look at the place where it had happened. Frost hung on the

Page 3: *Dad posed with the combine that shredded his shirt.*

tree branches. It was 7:30 and the school bus passed by on the gravel road going north, past the end of the driveway where I usually waited. I watched the bus go without me, wishing that I could go too, that somehow this could be a normal day. Searching the harvested rows of beans, I could find only traces of what had happened the day before. The old combine had been towed away, but I could see where Dad had died, for there were bits of his clothing, blue denim and white scraps of waffled long underwear, only a few icy drops of blood.

It was the combine, the same machine in the photograph, that killed him. The image was supposed to be a warning to us, not a prophecy. His death made that picture—the relief of a close call—too hard to look at anymore. It has stayed in the family album, in my mom's closet, until now.

Before the mower arrives, a field of timothy grass is green and alive, leaves waving in the morning breezes. Meadowlarks sing their early-summer songs, field mice are safe in their underground burrows, and a fox lurks by the fence line. Grasshoppers and crickets leap in the tall grass. A hawk stares from a nearby telephone pole.

The mower disturbs all life in the hayfield. The machine scatters wildlife and, if it lingers long, cuts it with scissorlike blades. Pheasants fly from the nest, leaving their young if they must. Timothy falls backward. It withers and wilts. The remaining stems bear a whitish cut-mark, slashed through.

In the aftermath (a word from the Old English, which means "after mowing") the whole field lies exposed. Naked. The short stubble lends no place to hide but under the windrow of timothy. There the crickets congregate, where they are easy prey for the meadowlark, the blackbird, and the sparrow. Some mice scurry in the cut portions of alfalfa and the hawk takes to the air, circling the field. Crows pick at the carcasses of other mice killed by the mower blades.

The grass appears dead too, but the roots are deep. The timothy will recover in time. New shoots arise from the wounded stems; the field will turn green again. Cows may graze there or the farmer will take a second cutting from the hayfield. In a dry year, the field might be plowed under and new crops planted, the old stalks nourishing the new. The cut timothy becomes hay, fodder for the livestock, sustaining them through the wintertime. Hay keeps traces of what it once was in the field, the faded green, the smell of summer, the strong perfume of memory.

Dad with little Larry, summer 1951.

My boyhood on the farm was a field of grass. My father and later my brother were cut down. The loss was deep, but a new growth of grass covers the field now. I can look out and see the meadowlarks on that hayfield again. The sun shines on the timothy once more.

I was never cut out to be a farmer. I never plowed a field; I never could figure out how to tell when a cow was done milking by feeling the pulse of milk through the rubber hoses; I never drove the tractor for mowing hay, for hauling manure, or for cultivating a field. I always let one of my brothers do those things; they wanted to drive tractors, and I did not. I did the hardest work, cleaning manure out of the gutters from the age of five, stacking the bales in the haymow, knocking down weeds with a weed cutter, picking cucumbers for making Gedney pickles, and digging up potatoes in the fall.

I did the manual labor and I learned enough to drive a tractor for some work, but I made sure not to learn too much. I mainly drove a tractor for raking or baling hay or for driving a full hayrack homeward. When I was fifteen years old, I would often drive the Allis Chalmers WD-45 tractor out of the farmyard, across the bridge over the county ditch, and out to the alfalfa hayfield to the east of the farmstead. My dad had already hooked up the side-delivery rake, which was fine with me; it would have taken me twice as long as anyone else. Every skilled job on the farm that was easy for the others was difficult for me, so I worked, providing the manual labor, hoeing cockleburs and milkweeds out of the rows of soybeans, carrying tall pails of milk to the milk-room cooler, pitching manure out of the calf-pens and the gutter, and feeding hay and silage to the cows. But I was good at making hay.

I didn't mind raking hay with the side rake, because all it required me to do was drive straight on the straightaways, slowing down on the four corners of the field in order to make the windrows easy to bale, always raking two swaths of mown hay into one windrow. I had it easy; I wasn't cutting with a scythe or even working with a horse-drawn rake. For early white settlers in Minnesota, haying meant weeks or months of daily handlabor. They needed to use their strong hands to make enough hay in the summer to last through the long winter months before spring, when the livestock could again be turned out to eat the grass in the pastures and meadows. Minnesota farmers throughout all generations made hay as an obligation; it was their duty. They cut and dried the grass, then tossed it onto haystacks and into haylofts for later feeding to the livestock. The labor of haying was never in vain, for hay was the food that kept their animals alive through the winter, but it was never a choice. Everyone needed it.

Even city people used to keep dried timothy-grass hay in their carriage houses and stables for the horses, but as we moved into the twentieth century, hay became less common. By 1970 there were more people living in the metropolitan Twin Cities area than there were cattle living on all outstate farms combined. In barely a century as a state, Minnesota's character had radically changed. Today many city dwellers have lost their connection to the country to the point that many now wonder, *What exactly is hay?*

Hay is humble stuff, just grass that has been cut and dried as fodder for cattle, horses, and sheep. But hay is not straw and should not be confused with straw. A farmer gets straw from the hollow stems of wheat or oats plants after the kernels have been threshed from the top of the stem; because straw is made of dry stalks, it has little nutritive value. Hay is green. Straw is bright, golden yellow. Hay is fed to the animals; straw is spread in animal pens to absorb manure and urine so as to keep beds dry in the cowbarns or sheds. Cows and horses eat hay but sleep on straw. A milkcow might sample a little clean, fresh straw put down for her bedding but certainly will not eat it after it has been trod under hoof.

Horses and cows eat grass in summer pastures. They need to eat fresh grass in order to have good health, but, in Minnesota, the pasture grass is not nutritious from the time it turns brown (after autumn frost kills it) until the springtime warmth and sunshine bring it back to life. When the grass is dead or covered with snow, then cows and horses have to eat the next best thing—dried-grass hay. Animals will eat it eagerly, for it has retained its nutritive value, especially its protein content.

Hay can also be made from the dried leaves and stems and reddish flowers of red clover plants, which grow to be about one and a half feet tall. Short white clover, famous for being the three- or four-leaf clover plant, also has become hay when it grows mixed with other types of grasses. Dairy farmers have come to love alfalfa hay most of all because it gives the most protein for cows to make milk. The leaves are especially tender and relished by milkcows, the stems are not tough to

Jan and Larry watching Dad unloading hay bales from our pickup into our hayloft with a John Deere elevator, June 1953.

digest, and the tiny purple, blue, and pink blossoms are lovely in the field and in the haystacks. Jersey and Holstein cows eat the whole plant in sun-dried alfalfa hay, so long as the hay is properly dried and promptly put into the loft of a barn or in a large stack outside.

However, if the cut grass has been rained on or not thoroughly dried, it can turn musty then moldy. Horses fall sick from eating moldy hay, and even cows can die from eating too much wet grass. The pasture grass ferments in their stomachs and can cause so much gas that they bloat. A farmer can save the cow if he diagnoses the problem quickly enough. He can put a stick in the cow's mouth to hold it open, allowing the gas to escape, but if the cow is lying on her side with her legs rigid, drastic intervention is needed. The farmer must take a knife or an instrument called a truncheon and poke a hole in the side of the cow to let the fermenting gasses out.

Once, as a boy, when one of our cows was on its flank in throes of agony, I watched my father take the barn knife we used to cut twine off haybales and puncture the cow's side. Gas came rushing out and digestive juices gurgled and flowed from the wound. It didn't help; before long that cow died in the cowyard near the pasture entrance. In winter, a dead cow would freeze; in summer it attracted flies that buzzed and circled the carcass. After a while, a truck came to pick it up, and it was turned into glue or dog food at the local rendering plant.

Farmers have always had to work hard to get the hay cut, dried, and harvested and into the hayloft in the barn as quickly as possible. It was always a race against time to get the hay harvested before it could be ruined by rainfall; thus the old phrase was born, "Make hay while the sun shines." It means to get the hay-work done before dark rain-clouds could arise and ruin the mown hay. It means to work as hard as you can for as long as you can before nightfall.

If this book succeeds, you will be able to feel the burning July sun and the noontime heat that made haying one of the hardest tasks of agriculture. You will find red clover plants in cut hay in a nearby meadow, take in the aroma and remember Shakespeare's line from *A Mid-*

summer Night's Dream: "Good hay, sweet hay."[1] Or perhaps, you will recall Robert Frost's assertion that the fragrance released by clover when cutting it with the "long scythe" made the work "the sweetest dream that labor knows." But the work itself was less than poetry, as anyone who has ever done the labor can attest. So I hope you will also be able to imagine the stifling heat in a hayloft and the cooling effects of an August breeze on the sweat-streaked brow.

Fill your lungs with the fragrance of hay. Hay permeates the past and the present with its gentle perfume. As Walt Whitman once wrote, "The familiar delicious perfume fills the barns and lanes"[2]; but perhaps the sweet smell of hay is less familiar than it once was. In 1991 the Smell and Taste Treatment and Research Foundation in Chicago conducted an investigation of the question: "What odor causes you to become nostalgic?" Researcher Alan Hirsch noted a powerful link between smells and memories. People born before 1930 associated their past with natural odors that they inhaled during childhood, such as hay, horses, meadows, pine trees, and sea air. Those born after 1930 tended to connect childhood recollections with the smell of a newly mown lawn rather than newly mown clover; with Pine-Sol rather than piney woods; with man-made scents like fresh plastic, scented markers, airplane fuel, VapoRub, Play-Doh, or Sweet Tarts rather than home-baked bread or wet clothes on a clothesline.[3]

Helen Keller appeared to be correct when she wrote: "Smell is a potent wizard that transports us across thousands of miles and all the years we have lived."[4] The fragrance of hay is a part of many pasts. No aroma on the farm brings back deeper memories than the smell of "good hay, sweet hay." The scent of red clover is the scent of haying heaven. Alfalfa's essence is somehow greener and heavier than clover, but still permeates the memories of those who fed it to cows and steers. Grass hay or marsh-grass hay leaves fewer traces in the nerves that govern smell but just as many connections to the work of gathering it and pitching it. Anyone who has ever lain down on his back on piled hay has kept traces of the feel of hay, of its aroma, its essence of work, its usefulness, of its very nature and quiet force.

The last words I ever said to my dad were: "It doesn't make any difference." I was saying good-night, standing in the doorway to the upstairs and both Mom and Dad were already in bed. All the lights were out except the hall light upstairs. I questioned whether I had been paid the full amount by the church for helping my brother Jeff with mowing the cemetery lawn; I thought I had earned a little more. He told me it was the right number, and I did the math in my head again. Jeff had indeed put in more hours, done more work. Conceding the point, I said to Dad, "It doesn't make any difference." My last words to him. Not the last thing I would have wanted to say, but words that resolved the matter, a gesture of concession. Maybe I couldn't ask for more than that.

We never said things like "I love you"; it wasn't the way we did things in our family. Our family had love—I worked with him and we had respect for each other and I never talked back to my father. That's what love consisted of on our farm. All we had ever really done together was to work together, in the barn, in the field—everything involved work, or developed a work ethic.

When I was a kid, my brother Larry and I were responsible for feeding the cows at an outside hay feeder, which was made out of an old straw-shed. It had two strands of an electric fence on one end, at the end where an outside stack of haybales stood. Larry was supposed to throw three bales over the wire fence, remove the twine strings, and spread the hay along the south side of the feeder, so that the cows and heifers could eat it after putting their heads between the partition boards.

I was supposed to throw three bales over the fence and then break up the slices of hay along the north side of the feeder. This I did until I began to feel that there was something unfair about the job. Feeling that I was too young and not strong enough to throw the bales over the wire fence, I committed a sin of convenience. Instead of feeding them three bales of hay, which was my chore, I climbed into the feeder after my brother was done with his part of the work, and I took half of the slices of hay that he had spread out on the south side and I threw them over to the north side. Because I was shunning my duties, the cows were not getting their full measure of hay.

It took only a few days for my father to notice that something was wrong. He could see that the cows were eating every bit of the hay in the feeder, rather than leaving some stems and rough parts. My father watched one evening as I did the chores, saw what I was doing, and then spoke to me that night.

I will never forget what he said to me. In a calm voice and with no anger, he told me that he was disappointed by my shirking of the responsibility of feeding the hay to the cows. He told me that we all had to work hard on our farm and that he knew that I would do the job the right way the next time. He said that he understood that throwing the bales over the fence was difficult for me and so he helped me with that part of the work the next night and for several nights afterward. That's how we measured love on our farm—by how hard you worked, how much time you put in. And, of course, it makes all the difference.

Work came first. Getting hay into the hayloft was a massive undertaking, for loose hay was measured in tons, not pounds. We had to lift tons of hay from the ground into the second-story hayloft. In the days of baled hay, one mechanical baler could produce more fully loaded hayracks in a day than could a whole family making loose hay. The baling machine in the meadow made it necessary to unload haybales into the hayloft as fast as was humanly possible. That meant that farmkids like us unloaded bales from the hayrack into the loft in the morning, in the noontime heat, or in sultry humid evenings. Many farmers put hay into the haymow no matter how hot it was, inside or out.

Each hayrack held over one hundred bales, and a bale elevator (or conveyor) lifted those bales into the hayloft with mechanical regularity, having no mercy and slowing down for nothing—not for heat,

Ane Marie, Larry, and Janice with squash at the homeplace, fall 1954.

not for humidity, not for time. The noise of the bale elevator went on and on and we tried to keep track of how many bales were still left to stack, but inevitably we would lose count. The first bales dropped off the top end of the elevator and fell a great distance to the floor of the hayloft. Some of thm broke apart upon impact and had to be pushed away from the landing zone. Hay leaves fluttered down, making a little pile of loose hay on the floor.

We put the bales in straight rows in order to get the maximum number in the hayloft space. It was a point of pride for farmers to operate an orderly farm, and a part of having such a farm required stacking the bales in the hayloft, rather than just letting them fall in a heap. The usual method in this procedure was to lay down a whole layer in rows running lengthwise east and west; the next layer would be laid in lengthwise columns north and south. Layer by layer the hayloft became filled.

The hardest work involved in filling a loft with baled hay was stacking the bales in the uppermost corners of the hayloft. The lower layers of bales involved much throwing and little lifting—after you mastered the talent of throwing bales into the right place. But when piling hay into the rafters near the ceiling, we had to throw the bales and then lift them into place—while the space became more constricted and the air was hot and close. I learned to walk with agility on top of rows of bales, and judging where to put each foot to avoid stepping into crevices swiftly became second nature.

Relief from the constant rush of falling bales came only when the screeching blades of the elevator and the electric motor stopped. When those of us in the haymow heard the elevator being shut off, we would think *Good!* knowing we were done with one more load. After this, it was a great feeling to get outside the hayloft door into the fresh air and sit down on the elevator and slide down on the smooth steel edges, made slippery by the polishing of thousands of bales that had gone up that elevator—being careful to duck through the haydoor and avoid any obstructions on the way down.

After filling the loft so full of hay that the green bales stuffed the gaping maw of the hayloft door, we could take pride in having gotten the hay safely into the red-painted barn. Red was the color of dairy barns, but the color green dominated all work with hay and haylofts. Farmers prided themselves on their good-looking hay; we wanted the hay to be as green as possible. Our cotton chore gloves were purely white or yellow when new, but all gained permanent tinges of green at the fingers, with bits of alfalfa leaves on the back of the hand. We even breathed in green alfalfa, getting a full nose from the green dust, and we blew green stuff from our nostrils at day's end. The right pant leg of the bale lifter became green with hay stains, and the fabric over the knee of the throwing leg wore out before that on the other knee. Dried green leaves from the hay gravitated into shoes and socks. They were unavoidable in haying time. They clung to you like memory.

The Catholic girls from my grade, even the ones I didn't know very well, all came to the funeral home for the viewing of Dad's body. I was glad they came, glad those beautiful girls cared. I was more of an observer than anything else that day, watching my mother's grief, my brothers' and sisters' grief—standing a bit outside of it all. I was trying to take it in but I don't remember much of the funeral time. Mostly what sticks with me is the image of his hands, folded in death, his hands, cracked from constant dipping in water while cleaning cows for milking and callused from twine digging into his fingers from one summer's three thousand haybales. What a bitter contrast to his smoothed-over, cosmetics-covered face.

I was fifteen, too young to really understand him; it seemed all he knew was work and all I knew *of* him was work. I worked with him carrying milk from the barn to the milk cooler. I didn't know what to say so we worked together in silence. Even on his fiftieth birthday that September, I had wanted to say "Happy Birthday," but I couldn't get the words out. Only later did I find out that he liked the fact that I was a good athlete, a good pole vaulter and a quick guard in basketball. He never saw me play, because he was working. At least he knew that I got good grades and did well at sports. At least we had worked together, every day, even if it was mostly in silence.

Life continued on the farm, and my eldest brother, Larry, though he had never wanted it, became a farmer. I did the same kind of work I had done before, manual tasks, but now I did it to please my mom and help my brother. Working with Larry, I learned to talk to him, something I had never known how to do with Dad. Larry had such a kind way of asking me to work; he said, "Do you want to grind the feed today?" I would only tell him "yes"; there was no avoidance of work anymore; there were no "no"s about farmwork anymore. Larry

was just eighteen, just out of high school, and had given up his dream of becoming a carpenter so that our family wouldn't lose the property my grandfather had purchased. For nearly sixteen years he managed to keep the farm go-

Mom and Dad with (L to R) Ane Marie, me, Larry, Dana, Jeffrey, Janice, north of the house, 1961.

ing, until he too was killed on that same family homeplace, less than a hundred yards from where our dad died.

This is a book of remembrance, a book tracing the role of haymaking in the lives of five farm families in different parts of Minnesota from the later Territorial period through the present. This could be merely a collection of facts about hay; but it is more, because hay meant life to livestock and, therefore, life to many Minnesotans. Each chapter follows a different family according to the changing methods of gathering hay during the decades of their lives, but this is not meant to be a detailed technical manual on haymaking, nor is it a book whose purpose is merely to remind or inform us of bygone haymaking methods of summers past. Instead I have tried to tell the stories of families on farms and how haying was part of the seasonal rhythms of their everyday lives, the larger rhythms of life and death. Those of us who grew up on Minnesota farms have only to count the number of farm deaths in our own communities to understand that every family will eventually suffer its own set of tragedies. This book is a tribute not only to those who lost their lives on farms but, also, to those who have endured despite those losses and continued to work their farmsteads.

My brother's story—the story of his life and death on our family farm—is the last of the five told here. I am a professor of history now, and some part of me needs to see what has happened to my family in terms of the history of our state, the trials of those who have gone before. I wrote this book to better remember my dad and my brother, so that they might be remembered by others. Wound around my memories of summers haying with my dad and my brothers are deeper threads of mourning. Danger, both natural and mechanical, is woven into the fabric of farmwork.

Sorrow is as common as labor on the farmstead. The dread-filled anticipation of the effort can be as heavy as the work in the field or hayloft. Whether grown on a back forty or wild on a prairie meadow, by July, fields of green hay—clover, alfalfa, timothy grass—are ready to be cut, raked, and gathered. When a field is ready, the horizon is usually hazy from the heat of the day and pollen thick on the air. It is the early afternoon, just after the main meal of the day. The stubble crunches, bending stubbornly underfoot. Most days nothing of note happens; you cut, you stack, you bale. But every day the fate of the farm hangs in the balance, and every day you're only one mistake away from disaster. You try not to think about it. Time to get to work.

"A Hope and a Future"[1]
HAYING WITH SCYTHE AND OXEN
Andrew Peterson, Swede,
Waconia, Carver County, 1862

In the summer of 1862, Andrew Peterson was caught amid three wars. Farthest away was the Civil War, in which Americans fought Americans; at that time the fighting was no closer than Tennessee. The Dakota War, on the other hand, came perilously near. That August, Dakota Indians stormed pioneer settlements along the nearby Minnesota River. What if the Indians advanced into Carver County and attacked his family or burned his house? The third war took place in his own soul—for as an immigrant Swede he was forced to decide: Should he help President Lincoln and his newly adopted nation fight its distant battles? Should he help his fellow white Minnesotans track down and defeat the Dakotas? Or should he stay home and protect his farmstead?

At age forty-three, he was not a young man with the stamina needed for marching and camp life, and Elsa was pregnant with their third child, due in December. Surely it was not right for him to leave his wife alone on their farm with two-year-old Ida, one-year-old George, and a baby on the way.[2] Besides, he had no argument with the rebels, no stake in the dispute between North and South. Nor did he have any quarrel with the Dakota people; he had bought his farm from a German immigrant who, as far as Peterson knew, had bought it from the government, not stolen it from the Indians.

19

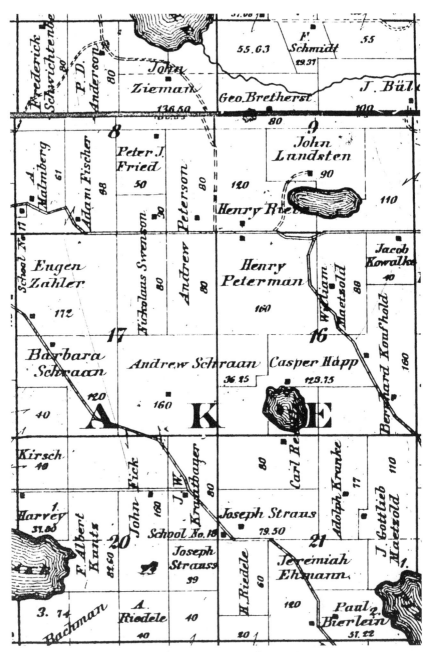

Andrew Peterson's two eighty-acre sections are shown near
the center of this detail of a map of Carver County, 1880.

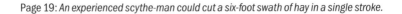

Page 19: *An experienced scythe-man could cut a six-foot swath of hay in a single stroke.*

He decided not to volunteer to fight the Confederates, but he was determined to defend himself and his family against the Dakotas. Should the warring Indians come near, he would try to get Elsa and his young children to safety, and he would defend his farm with his gun if necessary. At every turn it seemed Andrew Peterson might be asked to fight, when all he really wanted to do was tend to his family, till his fields, and make hay for his livestock.

My original hope was to find a Yankee or Old Stock American family to explore haymaking with hand tools, but the sources wouldn't allow it. Dependent upon farm diaries, I had to find one that told a complete story. My hopes were raised by the Frances M. Dyer diary that describes farmlife near Plainview in Wabasha County, but Dyer wrote as much about trying to find his free-ranging livestock as he did about making hay. Allen Dawley was a similar early disappointment. He would have been a good subject, because he joined the Greenwood Prairie Grange in 1874, was a schoolteacher off and on,

Andrew Peterson in front of the log cabin he built in 1855 near Waconia.
This photograph was taken late in his life.

and played some baseball in 1869, but his diary entries never supplied enough information to tell a story. I liked Levi Countryman because he had a wonderful name, but again, there wasn't enough to sustain a story. Diaries are often hit-or-miss affairs.

I was growing discouraged when I turned to the familiar story of a Swedish immigrant named Andrew Peterson. Not only were his diaries complete, but Peterson also had an arresting story. His life was based upon a few simple truths: he was a Baptist who worshiped his Lord according to the dictates of his conscience; he was a farmer who knew soil and weather intimately; and he loved growing apple trees and improving them by grafting hardy new shoots and buds into their trunks.

As a boy, Peterson had wanted land of his own, but he had no chance of finding it in Sweden. He was born in 1818 to tenant-farmer parents and spent his childhood working with his family as peasant laborers under obligation to the landowner. Families like his over-populated a land where poor soil yielded fewer and fewer opportunities to prosper. By the time he reached maturity, the best he could hope to do was work as a hired hand for his father. Thus, he left his homeland at age thirty-two, arriving in 1849 or 1850 as part of the first major wave of Swedish immigrants to the United States—each one hoping to get his own land and thereby improve his life. For Peterson, as for most of the others, attaining this goal was a slow process.

He spent at least three years farming as a hired hand, and on his own, in Iowa. The land prices there were too high for him to buy, but he began meeting and going to church with Swedish-American Baptists with similar ambitions. The preaching of Reverend Fredrik O. Nilsson appealed to him and before long he joined their ranks, accepting adult baptism in 1854. By this act, he cut his old ties to the Lutheran Church and renounced the worth of his infant baptism.[3] He followed Nilsson to the shores of Clearwater Lake (now Lake Waconia) in the Big Woods of Minnesota Territory in the spring of 1855. There, at age thirty-six, after five years in America, he finally had a chance to own land. He bought 160 acres of unimproved property in

Carver County from a German immigrant named Peter Fisher, who had bought it from the government. Now it was his. When Peterson first viewed his land, in late May, he saw nothing but green. Marshes merged with meadows of grass and wildflowers. He had hills and lowlands covered with tall oak and maple trees. Clearwater Lake lay only a mile away, and other small lakes and ponds were within two miles.

For all its natural beauty, it would take Peterson roughly twenty years of steady labor to carve out a farm amid the dense hardwood forests of Minnesota,[4] compared to only five years required to completely break the sod and cultivate a farmstead of the same acreage on the prairie grasslands. Yet Peterson felt comfortable surrounded by trees. For one thing, he knew he would always have enough firewood for the winter; those who farmed the prairies had to struggle to find winter fuel. He also would have enough wood to build the fences, large-haystack bases, and outbuildings he would need. The Swede turned his attention first to cutting logs with his ax and clearing enough ground on the south face of a hill for a cabin. Three fellow homesteaders helped him erect his first Minnesota home. After the cabin was built, Peterson "went and looked around to see" where he could make the hay he needed for his cow and her calf. Peterson would also need hay to cover the roof of the shed he had built for them until he could construct a larger, permanent barn.[5] Pastor Nilsson's nearby property had a grassy meadow. In exchange for the major portion of the hay he would cut there, Peterson helped the pastor erect his house and agreed to hold out enough hay for Nilsson's mare. On July 30, 1855, the end of the traditional haying month, Peterson sharpened his scythe and cut his first hay in Minnesota.

Waconia proved to be fertile ground for the growth not only of wild meadow hay and wheat but also of the beliefs of Swedish Baptists like Nilsson and Peterson. Lutheranism was Sweden's only official religion and law strictly forbade Swedes from following "unauthorized" churches. Punishments for disobeying this dictate included fines, prison terms, and, as in Nilsson's case, exile from the country. Here in Minnesota, those kinds of pressures were entirely removed. Whole communities of immigrants were free to establish settlements and build places of worship like the Scandia Baptist Church one mile west of Peterson's farm. His neighbors not only shared his Baptist beliefs but also his language, customs, and heritage, and this ethnic solidarity helped them claim a part of Carver County as their new Swedish America.[6]

The area was not entirely a "New Sweden" or a "New Scandinavia," however, for the rest of the county consisted mainly of German immigrants. While Peterson and other local Swedish farmers occasionally traded work with Germans, the language and cultural barriers were too high and they remained mostly isolated from one another, by choice. One of the Germans who lived only a mile away was Wendelin Grimm, who gained a place in Minnesota history by cultivating winter-hardy Grimm alfalfa. But the intervening mile was too swampy, the cultural divide too wide, and so they had little to do with each other.[7]

In addition to the neighboring Swedes, Peterson had relatives near Litchfield, forty-five miles to the west, and at Vasa, fifty miles southeast. Such family ties helped sustain him as he adjusted to the ways of his newly adopted country and weathered out the harsh extremes of Minnesota's seasons *and* the loneliness of his bachelor homestead.[8] Peterson's sturdy body could withstand the worst heat and cold that the Minnesota climate could throw at him. To the neighbors, he looked like a gnome, like a Swedish mountain troll, with a goatlike beard and a compact but powerful frame. For more than three years, he lived and worked alone, clearing his property and haying the Nilsson acreage. But soon, he found something better than the stays of work and distant relatives against his solitude: he met Elsa Anderson, also a Swedish immigrant, who lived with her widowed mother and her family in King Oscar's Town, ten miles east of Lake Waconia. After a brief courtship, they wed in September 1858. Andrew was nearly forty and Elsa twenty-three.[9]

In July 1862, as the haying season began, the atmosphere in Carver County, as everywhere else, was full of anxieties stirred by the Civil War. Bad news dominated: one part of the Union Army in a failed attempt to subdue Charleston had been forced to retreat at Secessionville, South Carolina (June 16), another part met defeat near Richmond, Virginia, in the Seven Days' Battles (June 25–July 1). President Lincoln immediately called for 300,000 more volunteers and a possible draft. Rumor said even older men like Peterson might be compelled into service, but, for now, farmers across the land, even where the war raged, had to gather in a hay crop. Unlike the South, where the animals could graze all year round and did not need dried fodder, the North would need hay-fed livestock for butchering in order to feed its growing army. As July began, he had no way of knowing that trouble was on the horizon as the Dakota tribe in Minnesota was growing restless after months of suffering from lack of food and justice. Andrew Peterson's thoughts were solely of cutting and gathering his hay. And he had to start the task the same way as always, by sharpening his scythe.

Peterson had bought a new scythe exactly a year earlier, for $1.25, at Warner's Store in Chaska.[10] It had a thin, crescent blade that fastened to its long, curved wooden handle, called the snath. The long sharp blade—distinguishing the scythe from the short-bladed reaping hook, otherwise known as a sickle[11]—attached to the handle at an angle so that the scythe-man could cut a wide swath of grass or grain (wheat, barley, and oats) with one sweeping stroke. Although the tool's name is properly pronounced "sīthe," Minnesotans typically dropped the lisping "th" and said "sigh"—as if the instrument itself were a deep breath of weariness or concern.

When Peterson began cutting hay on his south forty on July 15, his main worry was not about the Civil War or destitute Dakota people, but about getting enough hay from his meadowlands to last through the coming winter. He had reason to be anxious; the previous year he had been unable to make enough hay from his own land and had to buy some from his neighbors before the end of winter. This year, his sheds were full of livestock. He had three cows, each

Pair of working oxen about 1910.

with a calf, and one yearling heifer. There were also a pair of adult oxen and two young bulls (or bullocks), two ewes with lambs, and five pigs. He wanted to be sure to accumulate about two tons of hay per head of cattle so that he could feed it at a daily rate of about twenty pounds for each cow and between thirty and forty pounds per ox. Peterson would need to get at least fourteen tons if he was to keep his growing herd in dried fodder through a long winter.[12]

Cutting customarily took place during July because that was the month when the wild grasses were at the right height, and hay-making was fit into the time between cultivating other crops and the later onset of the grain harvest. His south forty-acre property, his best natural-grass hayfield, was just a short walk down the hill from the maple trees that surrounded his house.[13] And he would not be alone. His closest neighbors to the west, Andrew Hakanson and Nicholas Swenson, both Swedes, brought their scythes to help him. The men began to work together, cutting across the field, soon after sunrise on July 15; the grass cut best when the dew was still on the meadow.[14]

Good scythe-men cut with an almost metronomelike rhythm. The pattern of the scythe-man was simple: take a half-step and swing the scythe and then keep up a *step, swing, step, swing* rhythm—alternating the right foot forward and then left foot forward. Each half-step advance came at the same time that the worker swung the scythe back to its starting point. If the grass was thick, the mower had to take shorter steps. If the grass was thin, he could take longer steps forward. Mowing hay with a scythe was thus almost like a country dance: the worker had to feel his own rhythm and that of his partner—the swaying grass.[15]

In the times before machines replaced hand tools on the farm, a boy was considered a man when he could cut with a scythe. It took two adult hands to hold it, one at the snath's top and one at the perpendicular middle handle. Boys couldn't use it very well because the snath was long and unwieldy, and good mowers had both experience and strength. Without care or the necessary strength, a scythe also could be dangerous; inexperienced men occasionally slashed themselves in the foot or leg.

Each stroke had to be smooth in order to keep the blade of the scythe parallel to the ground and all the stalks of grass thereby cut at the same height. The scythe-man did not chop at the grass, using only his arms. Instead he rotated his body 180 degrees, a complete half-circle, twisting at the waist, keeping his arms still, swinging right to left and back again like a garden gate. The space the scythe covered in one long swinging stroke forward before it was swung back to its starting position, behind the scythe-man, was called the swath.[16] The width of one semicircular swath was about six feet, depending on the lengths of the man's arms and of the blade. The best mowers could "cut a wide swath" in an extremely straight path for as far as a hundred yards. Peterson, Hakanson, and Swenson, all experienced scythe-men, could make a six-foot-wide swath straight from one end of the meadow to the other. Efficiency and pride demanded that mowers leave no ridges of grass unmown between each mower's swath.

As the grass was sliced, it fell at the end of each forward stroke into an accumulating row—called a windrow—where the hay could dry before being gathered. Those who swung the scythe well kept the heel of the blade from hitting the ground and kept the point even with the heel, lest the tip cut into the soil and dull the edge of the blade. Thick meadow grasses were heavy enough that they fell easily in windrows, but thin grass scattered as it was cut and later had to be raked into rows.[17]

Experienced mowers made sure to keep the cutting edges of their scythes razorlike in order to slice through the grass stems smoothly and to reduce the amount of energy expended. The scythe-man typically kept a whetstone in the back pocket of his workpants, or in a cow's-horn container that hung from his belt. They paused about every fifteen minutes to sharpen the blade; the whetstone honing the cutting edge made a *ssshhh-shook* sound. Taking thirty seconds to a minute to apply the whetstone to the scythe actually reduced the total time required to cut the whole field, as well as affording a brief regular bit of rest. Mowing in July was hot work. Sweat would glisten on the mower's arms and dampen his shirt before the sun was even noon-high. Refreshment usually came from a clay jug of water that was wrapped in a wet cloth and placed in the shade of a nearby tree. The oaks and maples at the edges of meadows provided cool

Unidentified scythe-man about 1910.

places to rest on the hottest days, unlike farms on the open prairie that lacked shade trees.[18]

After Peterson cut the hay, he let it dry in place until the top leaves were well-wilted, then turned it over with a hayfork so that the sun could dry it further for an hour or two. The leaves of grass, being thinner than the stems, would dry more quickly in the sun than would the stems, making the leaves prone to shattering and falling off the stalk. The leaves would best dry without falling off if the farmer collected the cut grass into rounded mounds, about four feet by four feet, before it was totally dry. These mounds were sometimes called hayshocks but were more commonly known as haycocks.[19]

As with the cutting, Peterson usually had the help of neighbors in gathering the new-mown grass into haycocks. This exchange of labor was vital because he had no children old enough to work, and he had no spare money for hired hands. Work trading was common on the frontier in those days, when cash was scarce. That summer, a Swedish neighbor, Per Daniel Anderson, became Peterson's chief partner. He worked all of July 24 stacking hay at Peterson's place. Peterson returned the favor in equal measure the next day. In addition, two neighbor boys, Nicholas Swenson's son, Hans, and Albert Johnson, were valuable to Peterson for tasks such as cocking, stacking, and hauling the hay.[20]

Men, women, and children (as soon as they were eight to ten years old) all participated in making haycocks. In Peterson's day, farmers used handmade wooden hayforks with three tines, fashioned from tree branches of the proper shape, or a handheld rake, to pick up the hay from the windrows and pile it on the haycocks. They made a mound, packing it down by tapping the top with the fork. As more

---

To keep track of work his neighbors did for him and of the labor he performed for them, Peterson, like many other farmers, maintained a diary. He recorded exactly those chores that his Swedish neighbors helped him perform as well as the work with which he repaid them. He also made an account of what work he was owed and who owed it.

hay was added, the sides of the haycock were tapered and the top rounded. This semiconical shaping meant that only the outside of haycock would get wet; dew and rainwater would roll down the sides, letting the hay within dry, or "cure," safely and quickly—usually within several days. Grass cut at noon could be cocked by four o'clock in the afternoon. Then it could be gathered into large haystacks in the meadow and later hauled to the barn when the farmer found time.[21] The farmer would monitor the curing process by reaching his hand inside the haycock and feeling around. As hay cured it would produce its own moisture from within, a process called "sweating." When there was no more sweat, the hay was pronounced fully cured and ready to be put into a haystack. Peterson always asked for assistance when constructing big haystacks, for the task required several laborers.[22]

The first task in building a solid haystack was to prepare a strong foundation for it. Farmers used whatever was handy as "haystack bottoms." Peterson laid freshly cut logs, from trees cleared in the

Pioneer farmers stacking hay by carrying it on poles.

winter, on the ground.[23] Other farmers used brush or tree branches gathered from nearby for their bases. These foundations reduced spoilage because the hay did not touch the ground. In stacks built without such bases the bottom hay collected moisture and spoiled; in the winter, it froze to the soil, thus the farmer lost needed hay when he tried moving it to the barn for winter feeding.[24]

The next step involved carrying the small haycocks to the stack foundation on two long, smooth wooden poles. To make his poles, Peterson cut down smaller trees on his land and peeled off the bark. He sharpened the ends of the poles so that they could be readily pushed under each mound. It took two workers to move the collected haycocks. Each one put a pole underneath, gripped the poles, and hoisted, walking the hay quickly to the haystack and tossing it on. When the pile got too high for the pole carriers to lift their loads, they set the hay down next to the stack and used their wooden pitchforks to toss it to the top.[25] Because the poles did not pick up all the hay cleanly, children followed the carriers with rakes and gathered up the leftovers. Careful farmers such as Peterson believed in raking a meadow clean; leaving hay in the field was not considered to be "good practice," both because it was wasteful and because it might slow the continued growth of the meadow grass.[26]

Building haystacks was a vital skill on farms, and both Andrew and Elsa had learned the art of making them well. The couple made haystacks together in the years before their children were old enough to work or at peak times when his neighbors were too busy with their own labors to assist. On July 22, though she was four months pregnant, Elsa helped Andrew cock hay using the new handheld hayrake he had made that week. The rake had a long handle with a wooden bar fixed at the end. He had fitted the bar with downward-pointing dowels, forming teeth to help ease the task of getting hay to the haycock. Later, she helped him build haystacks.

An experienced haymaker such as Andrew Peterson combined efficiency of motion with strength in pitching hay. He jabbed the fork into the grassy mass, lifted it to shoulder level, and then launched it into the air, using his arms and the fork as a lever and his shoulders

as a fulcrum. The hay slid easily off the fork and flew up to the top of
the stack; its height, of course, depended on how high a man could
toss the hay.[27] Most stacks were between fifteen and twenty feet tall.
Andrew, with his greater strength, would pitch the hay up while Elsa
worked atop the stack, spreading the hay evenly with a pitchfork and
packing it down with her feet.[28] The goal was to keep the center of
the haystack from settling while making sure that the outer circum-
ference, which had not been packed by stepping on it, *did* settle, thus
making rainwater run off the stack rather than penetrate it.[29] It also
was vital that the completed haystack be circular and stand straight.
When the stack reached the desired height, Elsa finished it off with a
layer of swamp hay, forming a sloping top that shed water. This
process was called "topping," and it was critical that the last forkfuls
of hay be arranged so that the top of the stack formed a cone, allowing it to shed moisture effectively.[30]

In 1862 the Petersons built one large haystack on their south
forty, another on the north quarter-section of their land, a third beside the well, and a fourth near the ox stable and cowshed. Elsa and
Andrew usually made haystacks on calm days, because too much
wind blew the grass around, making it difficult to stack. They also
avoided making stacks on rainy days, of course, because water would
get into the mass of already cured hay and spoil it.[31] Tall haystacks
were part of the landscape not only in Carver County but all across
the farming areas of Minnesota. Weathered by sun, rain, and wind,
the once-fresh-cut, once-green grass looked gray on the surface of
the stack. The inner hay, however, still kept traces of summer prairie
grass in its aroma and maintained a faded green tint.

Peterson, like most farmers, put up split-rail fences around all his
haystacks to keep roaming oxen, cows, and pigs from devouring
them. He cut the rails from the trees on his land and split them himself. He made the fences high enough so that the tallest animals could
not jump over them and strong enough to prevent oxen from easily
breaking through them.[32] Animals were not the only hazard for
haystacks: parents ordered their children to *Stay off!* This was a rule
farmkids rarely violated, understanding the need to preserve the outside of the haystack, not to mention their own backsides.

Although haymaking was vitally important, Peterson never did it on Sundays. He believed in observing the Sabbath faithfully, and so even when he felt especially pressed by work, at whatever time of year, he let it go until Monday. The day of worship was a part of his regular rhythm; in the summer, it refreshed him for the upcoming week's haying or harvesting in his wheat field.

Peterson left big haystacks in the fields for two main reasons. First, the meadows were in lowlands, and getting the hay off the moist ground became easier to do once the ground had frozen. Second, once wheat harvesting began, he had to spend all his working time on that because it was his most valuable crop. (If his wheat was damaged by a hailstorm because he had been doing other tasks before cutting the

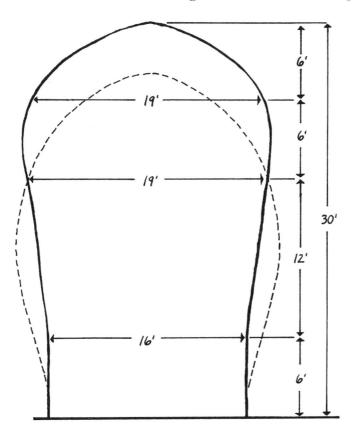

Cross section of a haystack. This diagram shows the shape of a stack built to withstand weathering. By keeping the center of the stack high and well packed while building it, most of the settling can be confined to the sides, resulting in water-shedding.

wheat, he had no one to blame except himself. Grass for hay could be cut during much of the summer and into September, but wheat had to be cut when it was ripe, in late July or early August.) Each haystack usually contained enough to fill two or three wagonloads; these were hauled to the barn in the fall or winter as the livestock needed.[33] Like all farmers, Peterson always had to be concerned about the weather during haying time. A violent thunderstorm could bring winds that blew the haycocks apart, soaking the hay to its core. When this happened, he had to spread the hay out to dry again and remake it into new haycocks.[34]

Andrew Peterson's haying season in 1862 lasted longer than most others. It began in June and continued, off and on, into September. In all, Peterson spent thirty days on haying work, more time than he expended on any other single farming task during that whole year. It wasn't weather or lack of help that unduly lengthened the haying season but a series of violent events that erupted as the summer drew to a close.

In August 1862, as Andrew Peterson was making hay, the entire southwestern Minnesota farming frontier exploded in the bloodshed of what is today known as the Dakota Conflict. Its most immediate cause was a shortage of food. No longer able to find wild game due to the encroachment of white settlers, and forced to wait for slow shipments of food, supplies, and money that had been pledged to them in several treaties, the Dakota people were desperate. As another winter approached, tribal leaders repeatedly asked the federal government to make good on its treaty guarantees. In August, a group of Dakotas came to the Lower Sioux Agency near Redwood Falls to claim their promised food. Although the food was on hand, storekeeper Andrew Myrick refused to distribute it to them on credit because the yearly government payment had not yet arrived. Myrick insulted them by telling their leaders, "If they are hungry, let them eat grass!"[35]

The spark was provided on August 17 in the Meeker County township of Acton when a group of hungry young Dakota men approached a farm to get some food and, after a dispute, murdered five settlers. Under intense pressure from other tribesmen, the Dakota leader Little Crow reluctantly agreed to call for a full-fledged uprising. He alone appeared to have the influence to lead his people in a battle to drive the whites out. Even though Little Crow knew the whites were too great in numbers for the Dakota to win, he saw no other way than war, since the killing had already begun. Having lost their freedom and their land, they felt they had nothing left to lose. The next day, warriors attacked the Lower Sioux Agency and killed Myrick, stuffing his mouth full of grass. The fighting escalated during the next three days as Dakota war parties struck farm families all along the Minnesota River valley.

Peterson and other residents of Waconia, a mere fifty miles from Acton, quickly got word of the attacks and believed that their settlement might be next, for warriors could reach it within "several hours" to "murder and burn."[36] Farmers in the open-prairie part of the state, where most of the early fighting was going on, had few trees to hide behind. Peterson realized that though the wooded hills surrounding his fields might provide a hiding place for himself and his family, they also could also be a place for the Dakota to lie in wait. Fear led them to flee "out on the island" in Lake Waconia on August 20, three days after hostilities erupted. They boated out to join other families in this "place of refuge," safe from any war party on foot. But they stayed only two days before going home. On the twenty-second, Peterson was obligated to fence in a haystack on a neighboring farm, lest it be eaten by free-ranging cattle. On August 23, Dakota warriors were attacking refugees and townspeople at New Ulm. While town defenders took up their guns and even pitchforks or packed hay into houses so they would burn quickly in case they were forced to torch their own homes, Peterson was busily engaged in making a haystack on his own north swamp. He may have brought a gun with him as insurance that day, for his wooden pitchfork would have made a poor weapon against an armed Dakota warrior. As he worked, he no doubt turned often to check the trees for signs of intruders.[37]

Fear turned to resolve as Peterson and his neighbors held a meeting and prepared to join the fight against the Dakotas. The men decided to get ready to march but waited for "more definite orders." Peterson and the others submitted to a doctor's examination to be pronounced "fit for war duty." On September 4, just thirty miles away in Hutchinson and forty miles away in Forest City, Little Crow's warriors were attacking more settlers' stockades. When news of these attacks reached Waconia the next day, another "Indian scare" went through the Swedish community. No one knew for sure which direction the Dakota forces had turned.[38]

Caught up in this anxious uncertainty, Peterson "did nothing" in the meadow for the next two days. The fearful proximity of the woods during the day was bad enough, but the sounds, real and imagined, that came from that direction at night were even worse. Nevertheless, Andrew and Elsa braved the woods, carried their belongings "out in the brush" to hide them, and got ready to "flee to St. Paul." However, no Indian attackers came, and within a few days the Petersons' fears subsided; with no reports of further attacks in their immediate area, the family quickly grew "more peaceful about the Indians." Unlike settlers in other places, such as St. Cloud, where farmers abandoned everything to seek shelter in the city, Peterson got back to haying on September 8, working with Per Daniel Anderson at Anderson's place.[39]

As Peterson worked to complete his summertime harvest, fearful settlers less than fifty miles away hid in hay to avoid being killed. Near St. Peter, Inga Johnson, a Swedish-American girl, crawled inside a haycock in a meadow, hoping that the warriors chasing her, who had already killed two family members, would not find her there. Her father returned, first rescuing a neighbor's wife and daughter from under other haycocks, before finding her. In another case, Charles Nelson "secreted a woman and child under a hay-stack" but could not find them when he came back for them.

See Lois Setterberg, "Carolina Svensdotter Johnson, 1828–1901" (family-history manuscript), 12. Copy in the author's possession. Nelson in Charles S. Bryant and Abel B. Munch, A History of the Great Massacre by the Sioux Indians in Minnesota *(Cincinnati: Rickey and Carroll, 1864), 146.*

The last great Dakota offensive came on September 23 at the Battle of Wood Lake, in which they suffered a major defeat.[40] Indians were captured, tried, and, by late December, were punished. A military court condemned 303 Dakotas to death for their part in the conflict, but President Lincoln commuted the sentences for all but thirty-eight who were hanged at Mankato on December 26, 1862. Although it appeared that the conflict was over, it had a solemn aftermath. As tales of atrocities, both true and exaggerated, spread from farm to farm and community to community, the fear continued long after the fighting ended. Estimates of white Minnesotans killed range from 450 to 800, to say nothing of the unknown number of Dakotas killed, making it the bloodiest of all Indian wars for the United States. A number of white Minnesotans moved away from the scene of the horror and never returned. Andrew and Elsa Peterson and their young family had not been harmed. Nor had anyone living in their vicinity. Warring Indians came no closer to Waconia than thirty miles away in the adjacent McLeod County. As the conflict died down, Peterson carried on with his farming, and Elsa gave birth to a baby boy, named John (called Axel), on December 16.

In 1862 Andrew Peterson had two oxen for plowing, pulling, and hauling. In addition, he was training two bullocks. He named one Brawn, for his strength and steadiness; the other was dubbed Berg for his mountainous size.[41] Peterson's oxen were effective draft animals and were well-trained and gentle. Both animals were much slower than horses, but they also had more power for pulling.

Plainly put, a steer is a bull that has been castrated. Bulls could be a nuisance if not a danger, for when they were intent on responding to cows in heat, only heaven could help anyone who got in the way. Frontier farmers thus routinely castrated very young male calves, transforming them into steers that would work as oxen. In Minnesota, farmers tended to have mixed-breed oxen. Some were from Shorthorn cattle crossed with another breed, such as Holstein or Hereford or Devon cattle. (Some farmers used female cows as draft animals, but male cattle were preferred because of their greater size and strength.)[42]

Like others who used oxen on the farm, Peterson trained the bullocks from an early age, beginning gently. After the animals turned six months old, the farmers would fasten two of them together with a small yoke so that they became used to being part of a team and to pulling a small cart. Ox trainers used a stick to tap an ox on the neck to teach it to turn. The ox would move away from the tapping and learn to turn to the left or the right. The training took place inside the fenced cowyard. Standing along the fence line, toward the center of the cowyard, an ox driver could keep them on track, teaching them also to respond to the voice commands "Gee" (turn right) and "Haw" (turn left). By these means, and with the early help of his neighbor Taylor August Johnson, Peterson readied his oxen to become steady draft animals. When the oxen were nearly two years old he accustomed them to wearing a full-size yoke.[43]

An experienced ox-driver, such as Peterson, had to communicate well because ox teams do not respond to reins or mouth-bits to guide them like horses. Ox trainers taught them to go forward or to back up by gestures, by waving a stick, or even by a well-timed grunt. In case none of the gentle means prevailed, he had an "ox-goad" (a club) to persuade his team to obey.[44] Peterson had made his own yoke, carving the large curved frame and the oxbows—each of which fit around and under an animal's neck—from the oak trees that grew on his property. The yoke had a large iron ring bolted to the middle of the frame to which he would attach a chain for hauling or a wagon or a plow.

The only haymaking task for which Peterson employed oxen was hauling hay out of the fields and to the barn. Peterson hauled a portion of the hay home in a wagon he had built in the fall of 1861.[45] Because the ordinary wagon box was too small to carry much hay, the farmer would remove it from the running gear (wheels and basic frame) and refit the wagon with a large wooden hayrack that extended out over the wheels both in front and on the sides. The hayrack held at least twice as much hay as an unmodified wagon. When the fall hay-hauling was done, the hayrack was taken off and stored. It could then be mounted on a sleigh or sled for winter hauling.[46]

In November came the cold weather that was just right for bringing hay in from the fields. The ground typically began freezing during the first week of that month. Soon after, the snows started, and the sleds were brought out. On November 18, 1862, Peterson yoked his oxen and had the team pull the sled out to get some swamp hay. There he used his wooden pitchfork to get the hay from the stack onto the hayrack and then drove the load home. The sled runners slid quickly and easily over the new snow. He unloaded the hay on a stack next to the cattle shed in order to meet the immediate needs of his livestock. On November 25, Per Daniel helped Peterson haul hay that was used to reinforce the hay roof of the cow barn.[47]

The most difficult moments of hauling hay in the late fall and wintertime came when the oxen had to pull the hayrack into an icy wind. To get a bit of warmth, the farmer would walk alongside the team as it went into the fields rather than sitting, elevated and exposed, on the hayrack. The exertion involved in clearing accumulated ice and snow away from the tops and sides of the haystacks warmed the farmer, and pitching hay onto the hayrack could bring sweat to his brow in even the coldest temperatures. Whatever the winter-day weather, Peterson had to walk rather than ride when directing the oxen home with a loaded sled, mainly to help ensure that the load stayed put. The oxen also had to work hard to break new paths through the snow as they hauled the hay. The work was not done until all the hay had been unloaded onto a stack near the cow barn and the ox shed.

While Peterson's oxen were needed for transporting loads of hay, they were most critical when it came to the heavy work of opening new farmland with a breaking plow. His team's strength and surefootedness were especially welcome when he plowed the hilly portions of his land. The chief complaint against oxen was that although they were powerful, they were slower than horses. Horses were especially well suited for working the mainly flat topography of southwestern Minnesota. Horsepower enabled a farmer to work a larger acreage than was possible with oxen, and newer machinery was designed to be pulled at horse speeds rather than at the slower speeds of oxen. However, this very slowness made them safer to handle. Farmers who used them experienced fewer accidents than those who worked with horses.[48] And though horses were better for all-around farmwork, they were too expensive for Peterson. An ox cost half as much as a horse—$50, rather than $100. An ox also was cheaper to keep than a horse, being able to function well on a diet of between thirty and forty pounds of either grass or hay per day. Horses, on the other hand, required oats along with hay in order to be able to pull with strength. Oxen did not get colic or other digestive problems from which horses often suffered. An ox could give a farmer many years of work, reaching its peak power from its sixth to its tenth year, and after that it could be fattened and sold for beef. Likewise, if a farmer got dangerously short on food, he could butcher an ox himself.[49]

Oxen, however, worked best in cooler weather and could not tolerate extremely hot summer days. In such climate, oxen would refuse to work, simply lying down until the temperature dropped to an acceptable level. When oxen got thirsty pulling a full hayrack in the summer, and smelled creek water nearby, they were known to take off running and pull the whole load into the water behind them.[50]

Andrew Peterson continued to farm without interruption to the end of the Civil War. He was concerned that, even though he was not a young man, he might have to join the Union Army. His younger brother Carl had gone off to war, and now the government had declared all men age twenty to forty-five eligible under the terms of the national draft law begun in March 1863. But in the end, he was officially excused from military service after being judged "too old to go to war."[51] At one point, he managed to contribute $40, a huge sum for him, toward clearing three local Swedes from conscription. (Those who were drafted and did not want to serve could pay $300 to cancel their obligation.)[52] The military necessity brought half of Minnesota's eligible men into the war, including some of Peterson's Swedish-American neighbors.[53] Without their help, his work became harder.

In February 1864, Nicholas Swenson let his boy, Hans, leave home to enlist. Peterson's workload increased considerably when in August 1864, amid haying and wheat harvesting, his valued working partner Per Daniel Anderson voluntarily joined the Union Army. Peterson helped keep his farm operating for the benefit, if not survival, of Anderson's wife and three little girls, until Per Daniel came home in 1865.[54] Having themselves had three children by December 1862, Andrew and Elsa eventually had six more. In all they had five sons and four daughters. As the boys grew up, they helped with all the farmwork, including haymaking. Elsa no longer had to labor in the fields.

It took nearly two decades for Peterson to clear and improve his land as well as to build permanent barns and a house. Year by year he increased the amount of land under cultivation, and year by year he built up his livestock and crops of oats and corn, expanded his garden of potatoes, squash, and melons. Until the early 1870s, he was chiefly a subsistence farmer whose main focus was on feeding and clothing his growing family. He didn't buy machinery until he had almost all of his 160 acres under cultivation.

Peterson cut hay with a scythe for the first eighteen years he farmed in Minnesota. In 1873 he bought a Wood's sickle-mower for

The N. C. Thompson Front-Cut Mower in the Franklin Steele, Jr. & Co.
Illustrated Catalog of Agricultural Implements about 1882.
The design is similar to the Wood's mower purchased by Andrew Peterson

$235 (at $25 down and the balance on an interest plan), but still had his oxen and so, for a few years, used them to pull it. He purchased a team of horses in 1877 and thereafter was able to make his hay faster. Peterson finally bought a horse-drawn hayrake in 1885.[55] Compared to Old Stock American farmers, modernization came late for him.

Peterson relied upon natural-grass hay from his meadows until he planted timothy grass in 1870, then clover by 1872. He later seeded alfalfa, which he called Lucerne clover, in 1886. He planted Grimm alfalfa, pioneered by his neighbor. He got more hay from the domesticated grasses and clover. However, he also had to do more work to get his hay crop into the barn because clover produced two crops in a summer, whereas the meadow grass and slough grass yielded only one. Alfalfa could give three cuttings in a season. So all in all, he got more hay and more work.[56]

After Peterson bought the sickle-mower, he felt the need to have a bigger and better barn to shelter his hay harvests. Accordingly, he built a barn twenty-seven feet wide by forty-two feet long, with a fieldstone foundation, in 1874. Seventeen of his neighbors came for the barn raising on June 25, and he filled the loft with hay in July. It was a "bank barn," one that was built into the earthen bank of a hill so that Peterson could drive his hay wagon into the upper-story loft and unload it on either side of the wagon. In December he moved the cows and sheep from their old straw-roofed sheds to the stone-walled bottom level of the new barn, and there they were more securely sheltered. Peterson and his sons fed hay to the livestock by simply throwing it down from the loft into the mangers. Ten years later he built a second barn, the same size as the first.[57]

Over the course of his long life on the farm, Peterson experienced almost every hardship known to farmers. He survived the grasshopper plague of August 1876, when "the air was so full of them that when you looked at the sun it looked as if it was snowing." He still got a hay crop that year because the grasshoppers generally preferred to eat grain and gardens rather than grass. Yet it was a smaller-than-normal crop and he had to buy hay during the winter.[58] When the grasshoppers swarmed again in July 1877, the Petersons even had to cut hay off the graveyard to get enough for the winter. Then there were the more routine natural setbacks. In 1878 "hail stones as large as goose eggs" hit his farm. During one July rainstorm in 1879 it poured all night, leaving water "standing on the meadows" and the hay floating around in the fields. And always there was the heat. On days when the summer sun and humidity combined to make the air so oppressive that Peterson and his sons could hardly breathe, he told his sons that they would not have to mow.[59]

Andrew Peterson was more than just a tough survivor. He became well-known regionally for his efforts to develop apple trees hardy enough to withstand Minnesota's winter climate. With 1,000 shoots from Iowa he started his orchard in 1856, the year after he first settled, and continued experimenting with apple grafts on his many trees.[60] He cultivated over 100 varieties of apples, especially Russian types like Hibernal, Anisim, Charmaloft, and Christmas, as well as Swedish apples. Peterson came to be thought of as one of the original Minnesotans, since he had settled here before statehood. He became a Minnesotan by clearing the trees from his farm, planting, tending, and harvesting his crops; by making hay for his livestock; by nurturing his apple orchard; and by raising his family here. He never returned to his native Sweden. Still, in a crucial sense, he can be said to have remained more a Swedish American than an American until 1889. That was the year he "met with an intense sorrow" when his "beloved daughter" Anna Isabelle died at age seventeen, the first of their children to pass away. "She was always so kind, quiet and tolerant," he wrote in his diary, "that our loss and sorrow is heart-rending to all of us. But in all our deep sorrow we have this comfort that she went away happy in her faith on her dear Savior Jesus Christ."[61] On Sunday, September 22, Anna was interred in the Scandia Baptist Church graveyard in Waconia. It is said that an immigrant's adopted land becomes his homeland when a loved one dies and is buried there. That was now certainly the case for Andrew and Elsa Peterson.

Peterson tended his farm right up until his death in 1898, at age seventy-nine. He died at the end of March, just before the apple blossoms bloomed. His closest neighbor, Nicholas Swenson, who helped him cut hay in 1862, wrote that Peterson was a "very God-fearing man" who "did his duties accurately and with good sense"; a "quiet man with an honest and good character, and nobody had any reason to complain about the way he took care of things."[62] Elsa lived on at the farm until her death, at age eighty-six, in 1922. The Petersons' children managed the place until 1926. Remarkably, none of the nine had children of their own, and so the Waconia farm passed into other hands.

Nevertheless, Peterson secured his place in Minnesota's history through his diary. In it is a record of all his days from the time he arrived in Waconia in 1855 until two days before his death. In 1939 it was donated, along with other Peterson family papers, to the Minnesota Historical Society by his family. During the late 1940s, while the great Swedish author Vilhelm Moberg was researching his four novels about Swedish immigrants in the state, he read the diary; and he drew upon it as he wrote the books *The Emigrants, Unto a Good Land, The Settlers,* and *The Last Letter Home.* Some believe that Moberg modeled the protagonist, Karl Oskar—the stubborn settler who remained on his farm to harvest hay during the Dakota War— on Andrew Peterson.[63]

I visited the Peterson farm site only once, on July 18, 1996—haying time. The forecast was for eighty-five-degree heat and plenty of humidity. Just after noon, I stood in the shade of a maple tree by the driveway and felt a cool breeze; but as soon as I got into the sun, it was *hot.* I looked down a slope across a small vale and hills to the next farmplace. I could smell clover, its big pink flowers in full bloom. I saw a couple of pig thistles in the horse pasture and a few Canadian thistles close to the wooden fence near the south barn. The stone foundation is cracked now and patched with concrete; the windows are almost all broken out, sashes rotting in the frames. The old red boards are sun-bleached and splintered, but on the north side of the barn, where the boards have fallen almost completely away, exposing the hand-hewn wooden beams that frame the haylofts, you can still see round bales, bound with twine, stacked to the rafters. The current owners also house a few horses there. The north barn still holds some of Peterson's old sleigh runners and other rusted-out equipment. Both barns are a story and a half in height, reminders of the labor that the Peterson put into carving his farm out of the Big Woods. He would hardly recognize it today. The land slopes down to Highway 5, a busy highway just south of the property. While the sun was still high, I drove that road back from Waconia toward home.

Andrew Peterson's south barn, Waconia, July 1996.

Our family farm-place, about five miles outside of Morgan, Minnesota, was squared and oriented according to compass points. The grove of trees on the north side protected us from winter winds; the barn was east of the house, so that the odor of manure from the cowyard would only rarely blow toward the house; the house faced south in order to get the sun in winter and the breezes in the summer; the gravel county road was just to the west of the farmstead, running straight north and south. I could always get my bearings on the farm.

But these directional pointers got all mixed up whenever I went to town, because Morgan was laid out at an angle. The town had been built according to the Chicago and Northwestern Railway tracks—which connected Sleepy Eye and Redwood Falls, running from the southeast to the northwest. Every street was straight in relation to the railroad tracks, and to one another, but the whole town lay kitty-corner with Main Street. All side streets ran from southwest to northeast, all cross streets southeast to northwest. No one could tell you where true north was except on clear nights when the North Star was out.

Despite this confusing layout, town life seemed like a paradise when I was a boy. As far as I could tell, the town kids (as we called them) had no work to do and could play all the time. They were good at basketball and baseball and other sports because they didn't have anything else to do with their time. I wanted to play all day instead of getting up early to work at baling hay or hoeing weeds out of the soybean fields. I wished then that I could be one of them, but once or twice, on our trips to town, the town kids taunted me for being a farmboy. They called me names—whether it was "hick" or "hayseed," "country boy" or whatever, I can't recall. What I remember is being teased because of where I came from. I felt ashamed of being from a farm when those town kids hollered those names at me.

So I tried to hide the fact that I came from a farm. Rather than wearing short-sleeve shirts when raking hay in the alfalfa field and getting a "farmer's tan," which only went from my hands to halfway up my upper arms, I took to wearing sleeveless T-shirts. The muscle shirt gave me a tan like the town kids'. I would even spend some time tanning my legs by wearing short pants and lying on a towel down by the apple trees, far out of sight of my parents and brothers and sisters.

I wanted to be in plenty of extracurricular activities in school, and so I played basketball and was on the track team. I wanted to play football when I reached junior high age; Rodney Harman, my history teacher and a coach, asked me to be the quarterback on the seventh-grade team. Hopeful, I asked my parents if I could go out for the team. Knowing that they needed help during the fall harvest, they said no. I found out later that Dad regretted having to say no and that if I had asked again, he would have told me to go ahead. But then, I took it as an absolute and never asked a second time. Farmwork had to come first, I figured, and for most of my upbringing, I resented that fact.

In retrospect, I have come to believe that I was better off having grown up in the country, in part because there were more temptations to do wrong in town. For lots of kids it seemed as hard to figure out the right thing to do in town as it was to find north. Eventually I did leave the farm. I went off to college and on with my life. But whenever I needed to get my bearings I went home and visited my family at the farm. My brother Larry would listen to my

The homeplace, five miles outside of Morgan, probably 1956 or '57.

problems and my dreams, and he would help me sort out which direction I should I follow.

My mother would listen, too, not giving much advice, showing her confidence in my judgment. Years later, after Dad and Larry died, Mom moved to Morgan. But true to her nature, she moved into a house located on a new street on the edge of town, the only street in town that runs straight and true, west to east. I can still get my bearings when I go home.

A Yorker's Sojourn in Minnesota

HAYING WITH HORSES

Oliver Perry Kysor, Old Stock American
Maine Township, Otter Tail County, 1883

In the coldest part of January 1883, Oliver Perry Kysor kept having dreams, dreams he couldn't shake. He had visions of his father's farm near the high headland in New York, the place called Kysor Hill ever since his family arrived there from Vermont in 1832.[1] That hill had been the center of his life since Perry, as he was known, was three years old. His father, Charles, had taught him and his five brothers how to be good farmers there, how to care for horses, and how to harvest wheat and make hay. In the dreams he also saw his mother, Sally Sweet Kysor, who had died in 1845, when he was sixteen and she just thirty-eight. The Kysor men laid her to rest in the cemetery at the foot of the family hill. Perry stayed and lived most of his life there, long enough to marry, long enough to lay in the cemetery one of his own sons, Gilbert, who died in 1861 at the age of seven.[2]

Eighteen eighty-three proved to be a year of dreams for fifty-three-year-old Perry Kysor and his family. He was in his first year of living out his dream of owning land and making a farmstead near the American frontier. Although he had owned a farm adjacent to his father's in New York, Perry had always wanted to see the West. Kenneth Kysor, great-grandson to Perry, still lives in Cattaraugus County, not far from the family hill, but he told me that "wanderlust" seems to run in his family.[3] However, the reasons are

49

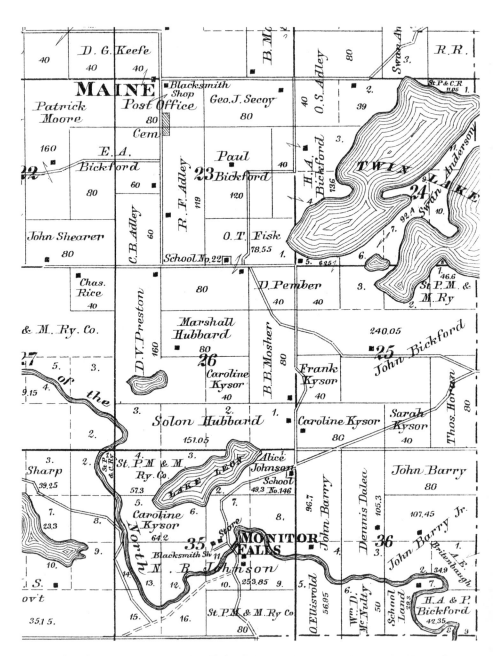

Oliver Perry Kysor's property, in his wife Caroline's name, is shown southwest and northeast of Lake Leon, in this detail of the plat of Maine Township, Otter Tail County, 1884.

Page 49: The Minneapolis Esterly Mower

probably more complex than that single word. The land records from Cattaraugus County reveal that an 1879 lawsuit over farm property and a mortgage was brought against Perry Kysor, his wife, Caroline, and six other individuals. He may have been forced to move, or bad feelings may have helped push him away to Pennsylvania. Perhaps Perry wanted to allow his sons to take over the farm at Kysor Hill, which might explain why he lingered in Pennsylvania in 1881, until a better opportunity emerged in Minnesota. When his in-laws, the Barret B. Mosher and Solon Hubbard families, invited Kysor to buy land in Minnesota next to the homesteads they had recently taken, he could not resist. He left two of his sons in charge of the farm on Kysor Hill and in 1882, together with Caroline (also fifty-three), and his five other children, most of them grown, journeyed to Maine Township, fifteen miles northwest of Fergus Falls.

Kysor family farm, as photographed by Flint & Nelson Portrait and View Artists, about 1900. Perry Kysor stands at center.

The Hubbards had been the first to migrate, in 1879. That year Solon Hubbard was forty-one, no longer young, but the arable land around Leon Township, in western New York State where he and other relatives lived, was nearly worn out. He sought a place where farming would not be such a struggle and found it in Otter Tail County, in western Minnesota; there the soil was wonderfully "new."[4] The land had everything a practical farmer could want—"prairie and timber, grassy meadows and beautiful lakes."[5] There was an abundance of white pine, maple, ash, and basswood trees. Importantly, there was "good soil, good water," and "splendid meadows" for making hay.[6] In all, the look of the area reminded him of western New York, except its low rolling hills were not as dramatic as those on the Allegheny Plateau, and many small lakes lay hidden in valleys instead of the Great Lake, Erie, that he knew.[7] The rolling hills provided assurance that the land had plenty of drainage so that crops would not be flooded out.[8] Further enhancing the agreeable environment was the fact that land cost less and property taxes were lighter than back home.[9]

But for Hubbard there was yet another comforting and attractive feature, one that had probably drawn Maine Township to his attention initially: it was a virtual replica of "York State," as upstate New York was known to its locals. In fact, although the place had been named, in 1871, by settlers from Maine, New York natives outnumbered all other groups by the mid-1880s.[10] There were numerous Norwegians in nearby Norwegian Grove and in Fergus Falls, but few Scandinavians lived in Maine Township.[11] Hubbard claimed homestead land near a small, spring-fed lake, which he promptly named Lake Leon, and soon helped his fellow Yorkers duplicate elements of the old-home environment as best they could. In 1881, he wrote his relatives B. B. Mosher and Kysor, back in Leon Township, of the abundant opportunities in Minnesota. At first Kysor was reluctant, but Mosher was convinced and, in time, helped persuade Kysor to move to Minnesota.

Settlement in Otter Tail County had been greatly facilitated by the opening in 1879 of train service between the Twin Cities and Fergus Falls on James J. Hill's St. Paul, Minneapolis, and Manitoba

Railway. The county's population immediately began to grow.[12] In 1883 the county's main commercial center, Fergus Falls, which had been incorporated as a village in 1872 and as a city nine years later, had a populace of six thousand. As a division headquarters of the St. Paul, Minneapolis, and Manitoba Railway, the city boasted a large locomotive roundhouse, a new opera house, a new schoolhouse, a fine courthouse, and eleven churches. Half its population came from "good society," meaning people from New England and other parts of the East. It was also a resort area, with the many area lakes in the so-called Park Region of Minnesota attracting tourists from Chicago, the Twin Cities, and even such faraway places as St. Louis. Even though Minnesota was not considered to be the wild frontier, the woods still held a multitude of black bears, which looked like "regular monsters" to some of these tourists.[13]

Why, as a couple in their midfifties, the Kysors decided to move their family west, leaving behind their ancestral home, will never be known for sure. The dominant problem in writing the history of private people from the nineteenth century remains the lack of complete records. Information too often is scattered or lost. Kenneth Kysor donated his great-grandfather's diary of 1883 to the Otter Tail County Historical Society in Fergus Falls, believing that the information would be of more benefit and interest to people in Minnesota than folks in New York. But Kenneth still has another of Perry Kysor's diaries, one from 1881. Without access to the diary, I have had to rely on Kenneth's description of what it contains. Perry lived in Derrick City, Pennsylvania, that year, where Kenneth says "he had a good business hauling well-drilling equipment and supplies."[14] If so, why did he leave Pennsylvania? Perhaps he preferred farming to business; perhaps it was his lifelong dream to live on the frontier, as Kenneth suggests. Perhaps it was just wanderlust.

Even where I have had access to the best records, a certain amount of speculation is necessary. Perry Kysor left behind one of the most complete diaries available on late-nineteenth-century Minnesota, but he wrote about what he did, not about his thoughts or his motivations. In his own reminiscences, titled *Time on My Hands*, Ken-

neth writes of this frustration, the gaps we all feel in our family his-
tories: "One wishes, at times, one could find all the answers to the
past. But then, as one considers all things, perhaps it is well the cur-
tain remains drawn."[15]

I have tried to make sense of Perry Kysor's life from what records
remain about the man.

The Kysors reached Maine Township in 1881 and Perry began
working to make his new Minnesota farm resemble the old homeplace, as much for his own sake as for Caroline's. They roughed it at
first; all eight squeezed into a log house comprised of four rooms and
a cellar that Perry built himself, but both he and Caroline agreed it
would be temporary. However, the log cabin was their home until
their new frame house was built in September and October of 1883.[16]
His farm also needed a substantial barn. On arriving, Kysor sheltered
his cows and horses in a large log stable, its roof made of wild hay
from the open prairie nearby.[17] The crude stable kept his livestock
warm enough those first winters, but it did not fit his idea of the kind
of barn a successful farmer should have.[18]

What he wanted was a substantial structure for housing livestock and storing hay, large enough for his new prairie homestead
and strong enough to meet his father's standards. Back in Leon
Township, when a barn was needed Charles Kysor was the man
called in to lay it out. He selected the heavy timbers and marked on
each the locations and sizes of the mortises and tenons. After he
completely planned the structure out, all the neighbor families came
for a barn raising. The men assembled the framing for the side walls
on the ground and then raised them vertically on the stone foundations. They framed the ends of the barn and put up cross-member
beams to connect the sides and ends. Then they fitted in the rafters
for the roof. Adjacent timbers were firmly attached with hard
wooden pins, pounded into holes that had been bored through the
mortises and tenons. When the framing was done, the host family
put tables out in the yard and set them with the abundance of food
the families had brought. The workers celebrated the barn raising

This famous photograph of a barn raising on the Rainy River shows how neighbors would turn out to help with construction.

in accordance with the temperance attitude of the farmer whose barn was being built.[19]

Not long after settling, Perry Kysor began working hard to cut enough timber to make beams for a real barn. In late September 1882, thirty of his neighbors helped get the frame up.[20] Kysor and his sons finished the interior work during the winter, laying the floor and erecting stall partitions in April. With this new barn, he no longer had to keep the hay in stacks outside the cowshed. It was a typical New York–style barn, with three bays: one held cow and horse stalls; another had an alleyway where the hay wagon could be driven for unloading; and the third bay held hay.[21] On the west side was a large door through which the men drove the hay wagon.[22]

His farm looked more complete with the new barn, but a New York–style farmstead, by custom, also needed an apple orchard. So Kysor planted apple trees, in the spring of 1883, not only for cider and for eating but also as a "source of profit and comfort."[23] Kysor kept up other traditions, too. He grew popping corn to cook in a woven wire popper over the evening fires. In New York it was considered "almost a daily dish," but in his first years in Minnesota he couldn't grow quite enough to make it an everyday treat. He also cut blocks of ice from Lake Leon, discovering that Minnesota ice in January was thicker than New York's, doubtless due to the "polar waves" of cold air descending from the north. Cutting household firewood and timbers for the barn left him with plenty of sawdust for covering the ice blocks in his icehouse.[24]

Not all of Perry and Caroline's children had made the move to Minnesota. The two oldest sons, Charles (thirty-one) and Willis (twenty-three), had gotten their own farms near Kysor Hill and had not wanted to leave their home state. Perry had given Willis an early legacy: his farm. For the five children who accompanied them west, the Kysors strove to maintain a sense of continuity. Perry tried to teach his two younger sons to be good farmers and his three daughters to become good farmers' wives. He wanted to instill in them principles of "wisdom, virtue, affection, industry," and truth, but at times he was not sure how well each child had absorbed those lessons. George, thirteen, got a "punishing with a ruler" from his teacher. One day in June, twenty-year-old Frank drank too much sweet cider and let the two horses, named Molly and Kit, run off, resulting in a broken harness and injuries to both animals, and considerable damage to the buggy. The girls appeared to do better, overall. Maud, at age eleven, proved to be competent at running the household when Caroline took sick in January. Eighteen-year-old Alice taught school part-time but also helped at home with various household tasks. And twenty-six-year-old Sarah, known as Satie, gave neighbor children music lessons in the Kysor house.[25]

The household included a hired man, Matthew Marsh, who had journeyed with the family to Minnesota to help establish their farmstead. A proven worker, Matthew, thirty-one, helped Kysor clear land, break ground, and complete the barn. Although he was not literate, Matthew was a good man to have on hand.[26]

Religious life for the Kysors remained quite the same as it had been back in New York. He and his Yorker neighbors and relatives were devout Seventh-Day Adventists. Numerous prayer meetings were held in the Kysor, Mosher, and Hubbard homes as well as in the

White School House (No. 22) and the Johnson School House (No. 146). Although they had not built a church, the families often heard guest pastors—including S. M. Bronson from nearby Evansville—preach at the White School House on Saturdays or Sundays.[27]

The Yorkers also frequently got together for social occasions. With their farms clustered closely, they were but a short buggy or sleigh ride apart. In late March 1883 the Kysors had the Hubbard children over to the house for singing, and the Hubbards, in turn, hosted a taffy pull. Also that month, the whole Kysor family watched their neighbors stage a play, *The Last Loaf,* at the White School House.[28] The Kysors held a neighborhood Fourth of July celebration in their grove of trees near the house. The summer of 1883 also brought the local families together for berry-picking excursions.[29]

The neighboring Bickford family, originally Vermonters, joined the Yorkers in worship and in fellowship. John Bickford and his five children especially came to rely on the Yorkers after his wife, Diantha, died of dropsy in December 1882. The night before her death, Diantha, blinded by her disease, had a dream about the glories of heaven. It had given her comfort and inspired John to bind the Gospel message closer to his heart.[30] John himself nearly joined her the following June when his colts reared up and knocked him down with their hooves. But just three days later he was able to attend a prayer meeting at the Kysors' home. Many of the prayers that day were offered in thanks for his recovery.[31]

Although Yorkers and other transplanted Old Stock Americans found conditions in Otter Tail County generally to their liking, they had to adjust to several powerful contrasts between their native ground and the new surroundings. For one thing, the geography was different. While the eastern portion of the county had plenty of "broad forests" like New York, it also had virgin prairie grasslands. Perry Kysor and his family claimed some of each, getting wild hay from the prairie and cutting the woodlots for fuel. Western Otter Tail County had even more prairie and bordered the tabletop flat Red River Valley, noted for wheat raising rather than mixed farming.[32]

Temperature extremes were also greater than in upstate New

York. There, winter temperatures rarely fell below zero, but January in Minnesota brought bitterly cold subzero days and nights that disrupted normal work patterns. One day when the thermometer hit minus thirty-nine, Perry and his sons did only basic barnyard chores rather than trying to haul hay. At times the wind "blew a perfect blizzard," and they stayed warm all day by the fire in their house, sheltered by a grove of trees.³³

Summers were hotter than in upper New York State. One day in 1883, the heat nearly gave George sunstroke, and a few July days were so hot that none of them could work outside. A severe midsummer rainstorm caused water to "run into the barn." The flies and mosquitoes were "very troublesome" as well, even though the Kysors no doubt knew about treating mosquito bites with applications of "spirits of hartshorn" (liquid ammonia).³⁴

Some things were the same, however: the seeds had to be planted, tended, and harvested. Kysor had plenty of help from his boys. In April, Frank grubbed out stumps and brush both on his father's property and on his own adjoining forty acres. Perry sowed the wheat and oats, starting on April 12. They planted peas on April 30, potatoes on May 1, corn—some on May 16, more on June 8—sweet corn on May 16, and cabbage on May 25. They also plowed some of their virgin prairie land and seeded it to beans.³⁵

Otter Tail County residents took pains to avoid traveling in winter storms. Everyone knew the sobering tale of Cassius Sherman, who had died in the "terrible blizzard of 1873." A Civil War veteran, Sherman put on the coat of his old blue uniform one January day and went to Maine village to get some cough medicine for his ailing mother. The weather was still fair when he headed back home, but a blinding snowstorm soon began. Sherman wearied and got lost in the drifting snow. Three months later, two boys spotted a blue coat sleeve sticking up out of a white snowdrift. Sherman's fate convinced locals then that no one should venture out alone during snowstorms.

See "Early Reminiscences of Otter Tail County," Fergus Falls Weekly Journal, July 12, 1929, Maine Township folder, Otter Tail County Historical Society, Fergus Falls, MN.

Kysor also required hay, of course, for his horses and cows. Unlike the solitary tasks of plowing and harrowing, haying involved the whole family.[36] By contrast, the wheat harvest and threshing included not only the family but also the whole community of nearby farmers. Kysor, his boys Frank and George, and his hired hand, Matthew, worked together to gather both "tame" hay, which was from grass planted by the farmer, and "wild" hay, which grew uncultivated in the meadows.[37]

Perry Kysor belonged to the first generation of American farmers to see machines as the means to successful, modern farming. For them, the most important farm implements for cutting were the mowing machine for grass and the reaper for wheat. These devices, invented in 1831, and in widespread use by the 1860s, allowed a farmer to cut ten times as much hay in one day as a man who used a scythe. Although Kysor managed to get along for several years by borrowing a neighbor's mechanical reaper for his wheat, he believed he had to have his own hay mower. He could understand why recently arrived Norwegian immigrant farmers nearer to Fergus Falls might still be cutting hay with scythes, for they had spent most of their money just getting to America. But only a bad manager would use hand tools for that task when he could afford this labor-saving machine developed by Yankee ingenuity.

Kysor also understood that the new immigrants had to rely on oxen as draft animals, not being able at first to afford horses and horse-powered equipment. Likewise, the first settlers to come from New York State to Otter Tail County in the early 1870s had oxen and continued to be "bullwhackers" until the railroad reached Fergus Falls, bringing more settlers, horses, and modern equipment to the area and transporting the wheat crop to the flour mills in Minneapolis.[38]

Kysor depended on his team of workhorses, Molly and Kit, to pull his farm machinery, and he provided them with just the kinds of feed they needed to perform their jobs well. Hay and pasture grass were basic foods for horses, while oats put power in their steps. Horses thrived best on hay made from timothy grass, because it re-

tained little dust and was the least likely to mold. In 1882 Kysor
planted timothy seed in a field a quarter mile from the house. Some
people called timothy "meadow cat's tail," because the seeds on top
grew brushy and resembled a cattail before they flowered. Many
farmers enjoyed nibbling on tender young timothy-grass shoots,
pulling the stem apart and eating the soft end; it tasted a bit like
brussels sprouts. The stereotype of a countryman with a timothy
stem in his mouth had basis in reality, for the hard part of the shoot
worked well as a toothpick.[39]

Kysor knew from experience that when the first dry spots appeared above the first joint on the stem, the timothy grass was ready
to be cut; this normally happened in late July. He also knew that if he
let the timothy grow too long, the stems would become "indigestible
woody fiber," as the moisture would be transferred from the leaves to
the seeds. Thus timothy grass that was cut too late lost much of its
nutritional value.[40]

That year, Kysor began cutting his timothy grass on July 30. He
started mowing in the morning, and the timothy, its stems twenty to
forty inches tall, fell in place after the quick cut with the mower. The
one he owned was called a five-foot mower, for it had a five-foot-long
cutter bar that sliced an equally wide swath through the grass. The
cutter bar, or sickle bar, bore twenty triangular cutting teeth. The

Cutting hay with a sickle bar mower, about 1910.

knife moved back and forth, each tooth of the reciprocating blade shearing off the stems in a saw-blade-like action. (Modern-day readers can understand the operation of the mower by picturing an electric knife or an electric hedge-trimmer, both of which use reciprocating blades.) The mower sawed off the grass stalk at about two inches from the ground. The cutting knife was protected from contact with rocks by triangular guards, which also channeled grass to the knife for cutting.

When Kysor mowed hay, he held the horses' reins and sat on a cast-iron seat that was part of the machine's heavy frame. He raised or lowered the cutter bar by means of a lifting lever at his right-hand side. To start mowing, he pressed this lever to drop the cutter bar to just an inch or two above ground level. Pushing a second lever engaged the gears so that the power could be directed from the turning of the wheels into the cutting motion of the cutter bar. This second lever served as a clutch that let the driver engage the cutter bar or take it out of gear when he needed to stop cutting. The mower was "ground-driven," so that when the wheels stopped, the mowing machine also stopped cutting.[41]

Perry Kysor, and other farmers of the 1880s, could figure on being able to cut nine-tenths of an acre of timothy grass per hour with the five-foot-wide swath of his horse-drawn mower, so long as everything worked properly. When he cut wild hay, however, he moved more slowly because he had to watch for glacially deposited stones and other obstructions. If the cutting bar hit a rock, several blade teeth could be bent or broken, requiring precious time and money for repairs.[42]

The whole process of cutting the hay began when the driver said "Giddap." The grass, as it was cut, fell down and backward, in a continuous thin layer over the ground. Because the blade extended to the right side of the mower, the driver naturally went around the field clockwise, with the cutter bar singing out its *click-click-click*.[43] In 1883, this was the new sound of mechanized agriculture, and the quick "clicking hum" of the mower soon replaced the "swish of the scythe."[44] When the operator stopped the machine and backed up to

In one of the last farming frontiers in Minnesota, in Beltrami County north of Bemidji, the Jacobson family had to make their hay the old-fashioned way. Ole Jacobson had come from Norway to establish a little farm place in the cutover pinelands near the village of Deer Lake. He had farmed by Starbuck and worked in a store in St. Paul, but he had put his hope into homesteading "up north." By 1908 he had built a log cabin and log sheds for his few livestock.

Jacobson gathered hay from the natural-hay meadow along the creek. It was lowland and quite swampy and he had to do all the work by hand. He had to work hard to cut the wild meadow grass with his hand-held scythe, and he could cut the grass by himself but needed help to get the hay home. He never had enough income for a hired man's help, so he depended upon his wife, Marit, and daughters, Edith and Agnes, to provide the efforts needed to haul home the hay. He had an infant son, Melvin, who would one day be his main worker, but that day was years away.

Haying had to be done in July, even though it was the "hottest time of the

Ole Jacobson family, about 1904, in Deer Lake, Minnesota, at the Lynx Post Office. (L to R) Henry Helgeson, Alice, Marie, Edith, Ole, Agnes, and Marit Jacobsen.

summer," because the grass near the creek was ready for cutting. It would not be much good for hay after the grass had gone to seed. Each of the four family members crossed the creek on a long log that reached from the home side of the creek to the meadow side.

Agnes and Edith used hayrakes, which their father had made by hand, to form haycocks. It was heavy work for the two barefoot girls, who wore workdresses in the July heat. It was "awfully hot down in the creek bottoms" where Father Ole used his pitchfork to collect the haycocks into small haystacks, ready for hauling across the creek where they would make a large haystack. In order to carry the loose hay in the small stack, Ole and Mama Marit used two poles. They put the poles under the haystacks and lifted the stack to haul it home. It was hot work in July and it was difficult physical work for Marit, who had always been sickly.

Ole led the way back home across the log over the creek. Marit followed behind, trying to balance her weight and the load and keep her feet beneath her, when it all went wrong. She slipped off the log and into the water, which was quite deep. The creek water was not deep enough to be over her head, so she could get out; but it was difficult because her dress and shoes became heavy in the water. Much of the hay also fell into the creek beside her and over her. The girls were scared.

Marit cried and wept from the frustration of falling off the log and for the effort of haying and homesteading. Those emotions and the helplessness of feeling so poor and so new to Minnesota, so far from the old home in Norway, overwhelmed her in that first year at Deer Lake. Young Agnes never forgot that day. Now in her nineties, she always remembered how she and her sister Edith comforted their Mama, who climbed out of the creek and once again collected and lifted her hay burden as they continued on their way home.

That winter Ole Jacobson fed wild hay that he and Marit had carried home by hand from the creek meadow. The cow and her calf ate the hay without a thought, for both would chew their cuds and digest the grass-hay later. Perhaps Ole remembered the haying time. On the way back to his log cabin, Ole would shiver as the January snow on his cheek intermingled with the close memories of Marit's summer tears, her fall into the creek water, and the July perspiration from his own brow—and maybe, just maybe, he wondered at the effort it took to live and farm in Minnesota.

From an interview with Agnes (Jacobson) Dickinson, Bemidji, MN, July 21, 1996.

turn a squared corner, the gear ratchets fell with a different *click-click* sound. After a farmer had cut an entire field, he then mowed the outside row of uncut grass, moving counter-clockwise.

The mowing-machine operator made sure that the horses pulled at a steady speed. The mower was inefficient when pulled too slowly, for it would break the grass off rather than slice it, in effect, strangling instead of cutting it. If the driver made the horses go too fast, the mower would not be able to cut all the grass in its path and the sickle bar would become clogged. When the mower was clogged, the driver had to stop, back up, and get up a little speed to clear the sickle bar. If too much grass was stuck in the blades, he had to push the excess away with a stick. Youngsters who operated a mower were always warned to keep their feet and fingers away from the sharp sickle blade when clearing it.[45]

Kysor's two horses had learned how to pull a mowing machine smoothly. Working horses developed a rhythm of their own, their heads nodding as they plodded steadily ahead. Cutting hay was vastly easier with a mechanical mower than with a scythe, and it was the horse that had to sweat now, more than the man.[46] If he relied on the scythe, the farmer had to cut grass when it was easiest to do so, that is, when the dew was still on the ground. However, that also meant that the grass required more time to dry. Grass cut more easily with a machine when it was dry, thus speeding up the curing time.[47]

While Kysor was cutting the timothy, Matthew Marsh raked the grass into haycocks with a hand rake. After putting the horses in the barn, Perry helped Matthew cock the hay in the cool of the evening.[48]

Obviously, Kysor had to be "weather wise" in deciding when to mow, rake, and stack hay. A farmer's efforts could go for naught if rain fell day after day after day, for if grass lay on the ground too long, it would become moldy and be unusable. The farmer had to be ready to work vigorously to get hay into the stack or the barn loft rapidly. He had to be alert to signs of impending rainfall and calculate when to cut the hay and allow enough time for it to cure so as to make fully nutritious feed for his stock.[49]

Timothy hay dried in just one day, so the following day, July 31, Matthew, George, and Perry were able to haul it home on a hayrack and put it into the barn. New timothy hay was just what the horses needed to propel them during the wheat harvest, which that year ran from August 7 through August 21. Of course, the horses also needed oats for strength, but they relished fresh timothy hay.[50]

Kysor borrowed John Bickford's reaper for cutting his wheat and oats. Wheat was Kysor's cash crop, while oats supplied feed for the horses. He planned the summer carefully, making sure to finish making and gathering the cultivated timothy hay before getting on with the grain harvest; thus neither crop would be spoiled by inattention.[51]

Once the wheat was safely cut and the bundles stacked for threshing, the Kysor men turned to the task of bringing in wild hay from the prairie. They got the prairie hay from open land that the government had granted to the railroad but that no one had yet bought. The prairie grasses were there for the taking by those willing to put forth the effort to cut it, rake it, load it, and haul it home. Kysor had to gather wild hay, because he did not have enough open meadowland on his property to cover all his hay needs. He used some of his land as animal pasture, but it could not be used for both pasture and for hay because the animals ate the grass down too close to the ground, and it could not grow back quickly enough to provide hay.

Out on the prairie, about two miles from the farmstead, the family used a large tent as their base of operations. Perry had made the tent's ridgepole himself, and he was satisfied that the tent could stand up to any rain and wind, save for a cyclone. The tent provided cooling shade for the workers on the treeless prairie. It also gave the Kysor men a place to eat and rest and to keep their food and tools. Perry kept an ample stock of such staples as flour, sugar, salt, and coffee, supplemented with vegetables and meat from the family larder. It was also where the grindstone used to sharpen the mowing machine's sickle blade was sheltered.[52]

Kysor's sons George and Frank did much of the mowing. Although George was old enough to operate the machine, he did cut his fingers on August 29 while sharpening the sickle blades. The injury didn't slow him, though; he was back haying the next day. Like all farmers, Perry Kysor and his boys had to be concerned about the potential for accidents. He insisted that the utmost care be taken during mowing, especially in sharpening the machine blades. Just a week before George was cut, an accident took place in the Red River valley that doubtless chastened farmers statewide. William Carson, a Polk County farmer, was driving his mowing machine, holding his three-year-old son in his arms, when the horses "became frightened and gave a sudden start, throwing the boy from his father's arms in front of the cutter." Carson tried to save his son and fell into the path of the mower, which "ran over both" of them. The child was "cut to death," and the father, "horribly mangled," also died. Carson's wife was so "badly shocked" that it was feared she would "never recover." In its report on this tragic incident, the Fergus Falls newspaper noted that more than two hundred Americans died each year in mower and reaper accidents.[53]

But even if some accidents could perhaps be avoided, there was no getting around the sweaty work of haymaking. When temperatures climbed into the high eighties or low nineties, the horses pulling the mower began dripping sweat down their sides and flanks. The man sitting on the mower had the sun in his eyes a third of the time as he went around the field, but the heat most directly affected the rakers and pitchers. Whenever there was no breeze, common flies and horseflies buzzed around both man and horse team. Sweat ran freely down workers' faces and they frequently paused to mop their brows with sleeves or handkerchiefs. On the hottest days, the Kysors all kept out of the sun when possible, but if they had to get the hay-work completed that day, they might, as the farmer's manuals advised, put wet leaves from prairie weeds inside the headbands of their straw hats to prevent sunstroke and heat headaches.

In such unrelenting heat, dehydration was a constant danger. The Kysors brought barrels of water from the well at home to the tent, most of it for the horses. Perry and the boys made daily trips from home to the prairie and had the horses haul the water barrels in the rear of the wagon. It is likely that they brought out to the harvest fields large chunks of ice from their icehouse, setting it in a covered, twelve-quart tin pail or an earthenware jug under the shade of the tent. The ice would then slowly melt for six to ten hours, providing enough cold water for a whole day. When a worker became over-heated, the general rule was first to "cool the hands and face, and hold a lump of ice in the mouth."[54]

Harvesting the wild prairie hay began on August 22 and ended on September 7. All the men in the family, and even the neighbor boy, Clint Hubbard, pitched in. Kysor rotated Frank, George, Matthew Marsh, and Clint from mowing to raking to pitching to stacking to hauling the hay to keep them from tiring of any one task. At night, Perry took the horses back to the barn and slept at home, but he left the boys to watch over the tools and supplies in the tent. The boys didn't mind sleeping in the tent amid cool night breezes.[55]

At the midpoint of prairie harvesting that year, the whole Kysor family had a picnic in the tent to celebrate getting so much hay successfully cut and dried. When the meal was over everyone went back to the house except Perry, who stayed to rake and mow, until the boys rejoined him in the late evening.[56]

Man mowing hay with a side-mower, similar to the one used by Perry Kysor, 1911. The horses are draped in fly-nets as protection from mosquitoes and horseflies.

The Kysor women were essential at all times in supporting the men in their labors. When the work went on near home, the men of course ate in the house. Caroline cooked the meals, and the girls set and cleared the table and washed the dishes and utensils. The women stoked the fire for baking, churned the butter, and prepared the apple pies in season. When the men were haying in fields distant from the house, the women and children brought food and drink out to them. Country people know that haymakers appreciate and even need flavored drinks while out harvesting. A sweating worker might tire of plain water, so many people made "switchel," the summertime drink that could restore a haymaker's energy even on the most sweltering of days. The basic recipe called for combining a cup of maple syrup, a cup of apple cider vinegar, and a half cup of light molasses, and stirring it all into a quart of cold water.[57] After that a key flavor ingredient—one tablespoon of ground ginger—was added; ginger contributed to the overall tonic effect and maintained the "proper amount of internal heat." Switchel thus quenched thirst while elevating blood-sugar levels. Moreover, the acid in the vinegar had a cooling effect and gave the concoction some zip.[58]

After the prairie haymaking was done, Kysor harvested some tame Hungarian grass that he had planted the year before. Also known as foxtail millet, it ranged from two and one-half to five feet tall. It made better hay than the wild prairie grasses but wasn't as good as the timothy. Kysor used it mainly to feed his cattle.[59]

The making of large haystacks had not changed from the 1860s, when Andrew Peterson built them near Lake Waconia. Like Peterson and countless others, the Kysors made the stacks in the field and left them there until they had enough time to haul them home. A key element in making a haystack was giving it the right shape. No matter the specific dimensions of a given stack, farmers molded it to resemble a hen's egg, with the "small end up," so that it would shed rain as well as a good roof would. Also, the base was smaller in diameter than that part of the stack just a few feet above. Thus after completing a large stack, a man could lie down at its base and be sheltered from sun and shower by a bulge of hay directly above.[60]

Clearly, horses made certain aspects of hay-work easier in the time of oxen-powered labor. Nonetheless, farmers still had to use their combined brute strength to get the hay from field to barn. In those days, pitching hay by pitchfork was termed the "armstrong" method, for self-evident reasons.[61] The pitchfork had three metal tines that were widely enough spaced so that the tool could jab into a haycock easily and easily release the hay once it had been tossed onto the hayrack or haystack. A pitchfork's smooth wooden handle raised no blisters on the already callused hands of Perry, Frank, and George Kysor. Mature farm workers such as Perry and Frank "manipulated the pitchfork with the expertise of experience," and George, though still young, was almost as good.[62]

Prudent farmers did not the leave pitchforks outside because rain and dew would cause the wooden handles to crack and deteriorate. The careful worker also dragged his pitchfork behind him as he moved across the hay fields since that took less effort than carrying it over his shoulder, and pulling it across the ground reduced the possibilities of tripping in hay stubble and perhaps falling forward onto the pitchfork.[63]

When the Kysors brought the wild hay home from the prairie, they pitched it from the field stack onto a homemade hayrack they had built from boards cut at a local lumber mill; the wood was from trees on their farm. They placed the rack atop their buckboard wagon's running gear. His small rack, which probably weighed about a hundred pounds, could haul home enough hay to see the livestock through the rest of September and the fall.[64] In late November and early December, when they needed to bring in greater amounts, they borrowed nearby neighbor D. E. Pember's larger hayrack. After sufficient snow had fallen, Kysor, like Andrew Peterson, put his own hayrack atop a sleigh and used it to get the remaining hay home from the prairie. The Kysors, having harvested more hay there than they needed, intended to sell any excess to local farmers who ran out during the winter.[65]

By the time the Kysors did their late-autumn hauling, the color of the haystacks had turned from grass-green to gray. The stacks had shed the summer and autumn rains, but the moisture and subsequent drying in the sun had weathered their outsides. As Perry and the boys pitched the hay to the rack, they uncovered green, and the aroma that was released likely reminded them of the summer days back when they had made the stacks.

Waiting until fall and early winter—when the thunderstorm season was safely over—to move hay into the barn meant not having to worry about the hazard of lightning strikes. Because Kysor's barn was the highest point of his farmstead, it had the greatest chance of being hit by lightning. If it were struck by lightning when filled with hay, it would be all the more combustible; and a fire that reduced the barn to ashes would, obviously, leave the livestock without either food or shelter. Not surprisingly, lightning-rod salesmen preyed upon farmers' fears of such occurrences. One such salesman came to Maine Township in July 1883, promising that his lightning rods would "catch the whole business" of electricity from out of the air and "return it to the Heavens again."[66] Some farmers wondered whether the devices actually attracted lightning rather than repelled it. In any

---

Before the turn of the twentieth century, traveling lightning-rod salesmen were common in rural Minnesota. Local farm-equipment dealers also carried lightning rods as standard items from 1880 through the 1920s, the era when the vast majority of Minnesota's barns were built. In Minneapolis, the M. Townsley & Sons Company promoted itself as general agents for the Omaha Lightning Rod and Electric Company's pure-copper-cable lightning rods in the May 15, 1905, issue of *The Farmer*.

Even today, if you look carefully at old barns and farm buildings, you can spot lightning rods on the rooftops as protection against electrical storms. They're fairly simple devices, just a rod mounted on the peak of a roofline with a conductive-copper cable extending down to an anchoring pole, which extends about three feet into the ground at the base of the building. Simple as they are, lightning rods often saved barns or homes from burning to the ground. When lightning hits a modern home without a lightning rod, the current will travel through the wiring into the main breaker box or down water pipes. When fuse boxes explode, a home can burn quickly.

event, Kysor did not have enough money to buy lightning rods that year; maybe he would the following year.

In all, the Kysor family had a good season in 1883, with a barn full of hay and additional haystacks out on the prairie. As they hauled that hay home in December, they might have looked over their shoulders at the bare prairie where they had worked so hard in August and September. Looking at the setting sun, they could not have missed observing a strange, dark-red "blood afterglow" in the sky, and Perry Kysor might have wondered what the brilliant sunsets might mean. He likely didn't know it at the time, but the red evening skies, prominent all across North America and visible as far away as England, came from volcanic dust spread in the atmosphere by the enormous eruption on the Pacific island of Krakatau in August, half a world away, while he had been haymaking.[67]

The transplanted Yorker family of Perry Kysor succeeded in bringing their culture and customs to Otter Tail County for but a single generation: all the Kysor children eventually left the Minnesota farm. Frank returned to New York; Sarah married and moved to Fergus Falls; Alice wed and went to California; Maud married a local man, Harry Pettit, and in time resettled in Wayzata, west of Minneapolis. George stayed around the area the longest, purchasing the general store next to Phelps Mill, after marrying into the family that ran the mill. He operated the store from 1899 to 1908, when he sold it to a cooperative and moved near Maud in Wayzata.[68]

Sorrows came to the Kysors several times after 1883 and may have contributed to the desire of the younger generation to depart. In 1887 cousin Franklin, twenty-nine years old, came out from New York to visit and work for the summer, eager to see his cousins and to observe what Minnesota was like. Employed as a hired hand by neighbor H. T. Putnam, Franklin went out into the pasture to bring home the cows early on a July morning. A forenoon thunderstorm arose, and the Putnam family grew alarmed when Franklin did not return home after a reasonable time. One of the children ventured cautiously out to the pasture to find that Franklin had been struck and killed by a bolt of lightning. About twenty feet from him were two cows, also dead. Uncle Perry arranged for the burial of his nephew in the local cemetery; he was the first Kysor to die in Minnesota.[69] Given his devout outlook, perhaps Perry eventually found peace, as Job of the Old Testament did, by recalling that "the Lord gave and the Lord has taken away; blessed be the name of the Lord."

Such an attitude toward suffering might have helped the Kysors cope with subsequent deaths in the family. As noted, Maud married locally; for a time, the couple lived adjacent to Phelps Mill. They had two daughters; both children died of scarlet fever in the same year. Maud and Harry soon relocated to Wayzata, never to return.[70]

When Perry Kysor was too old to farm, he and Caroline bought a house in the village beside Phelps Mill. Caroline, by now known to family and neighbors alike as "Grandma Kysor," died from "stomach trouble" in 1910. She was eighty-one. Willis came out for the funeral, but Charles and Frank were unable to make the trip.[71] His wife's passing left Perry alone, and he was too old to carry on by himself. He moved to Wayzata to be near Maud, George, and their families. Early in the summer of 1912 he fell ill, and on July 30, as haying season was getting under way, he died. His family arranged for his body to be taken back to Maine Township for burial. The pallbearers included H. A. Bickford, Israel Cameron, and D. E. Pember. They remembered him as "a Christian man, a member of the Advent Church, a good neighbor and friend."[72] After Perry died, no Kysor lived on the farm by Lake Leon.

The Kysors' faith had done much to sustain them during their three decades in Minnesota. They saw their time there as simply part of an earthly pilgrimage whose radiant goal was an afterlife spent among the citizens of heaven. This idea took many forms but perhaps none so vivid as that expressed by the eighty-one-year-old Charles D. Kysor while on a visit to his son's new farm during that year of dreams, 1883. After reciting his evening prayers and going to sleep, he awakened at midnight after a dream in which the bedroom was flooded with a "light different from any thing he had ever seen." He believed that the glow was reflected from the "Eternal City" in heaven, and later testified at a prayer meeting that he had been vouchsafed a glimpse of the Celestial City as he neared the end of his time on earth.[73]

Today, apart from their side-by-side graves in Silent Vale cemetery in the village of Maine, there are few remaining earthly signs of Perry and Caroline Kysor's long sojourn in the hilly countryside around Maine Township.[74] Their first farmstead still has open grassy fields and beautiful woods, but neither house remains standing, and the barn is gone. Yet traces of the legacy of the New York farmers who settled near Lake Leon can still be found. The most prominent reminder is the Phelps flour mill, built alongside the Otter Tail River in 1889.[75]

Each of the past two autumns, I have taken my family on a Fall Leaves Tour from Barnesville past Maplewood State Park and through the Phelps Mill area. We have picnicked at the Phelps Mill Park and toured the flour mill. At dusk in October, the mill's dark corners and four-story height suggest an earlier time in a way that even my children could feel. There is, of course, one further, highly valuable trace: the diary that Oliver Perry Kysor kept for the year of 1883. There he left what proved to be an enduring record of a year of homesteading, haying, and harvesting in Maine Township, enough to piece together a fragment of his life.

Phelps Mill as it appeared in the 1980s.

One week after President John F. Kennedy was shot in Dallas, I almost died. I was ten years old, and my brothers and sisters and I were swinging on a rope that hung down from the rafters of our barn's haymow. Over the course of the winter, as we threw the haybales down the chute to feed the cows and calves in the barn below, we cleared out enough space for swinging on the long rope.

During times of putting up hay, a farmer would jerk on this rope to release the latch on the hay carrier, dropping loose hay from the sling to the floor, but to us, that rope was more than just a tool for fodder storage: it was the source of hours of fun. Larry and Jeff and I would build a takeoff platform of bales on the east and a landing platform on the west. The intervening area could be a valley, a canyon, a river that had to be crossed by swinging on the rope. We might chase each other from side to side or just see how far we could fly.

My third grade school picture, age nine.

On the evening of Saturday, November 30, 1963, I clung to the rope as it took me over a valley—the valley of the shadow of death. President Kennedy had been buried the previous Monday—no school nationwide as our family joined countless others in watching the funeral on TV. Images of Jackie Kennedy kissing the coffin merged with little John Jr. saluting the funeral procession and the riderless horse with the empty boots put in the stirrups. I remember Kennedy's funeral better than I remember the rope swinging that evening. The only impression that I can now recall is that of landing on the bales and feeling something in my side give. I didn't tell anyone.

My brother Larry and I slept in the same bed that night, as we had for years. I kept to my half and he kept to his, neither ever crossed the line. I fell easily to sleep, but awoke in the middle of the night in agony. My stomach was swollen to the size of a basketball, the skin tight. What I didn't know at the time was that the little give in my side was my appendix rupturing, and now the bile had spread inside my body cavity around my intestines. That fluid was like poison, infecting the peritoneal tissues lining my insides; my body had begun to swell.

In the weeks before, I'd had no obvious symptoms of approaching appendicitis, just some vague stomachaches in early November that caused me to miss a couple days of school. But I was a slender-waisted kid, strong from lifting bales and doing chores; a little stomach distress was a minor irritation. When my appendix broke, the pressure on it was finally released, so that I actually felt better that night than I had all month.

But now I was full of fever, groaning. Larry woke to my thrashing and shook me, trying to get me out of the misery. Unable to do so, he went downstairs and wakened my mom and dad. My folks put me in the backseat of our 1957 Mercury where I lay, writhing, for a long eight-mile drive to the Redwood Falls Hospital. Our physician, Dr. Ceplecha, told the nurses to put tubes down my nostrils to my stomach and drain the infectious fluids. The last thing I remember is the nurses trying once with large tubes and getting partway down, scraping all the way. When they got stuck, they pulled those out, scraping all the way back, then got smaller, child-size tubes that worked. I passed out.

The doctors could not operate for appendicitis because there wasn't an appendix anymore. Not knowing what to do, they called in Dr. Black from New Ulm. He delivered massive doses of antibiotics intravenously to kill the infection. I can only imagine what my mom and dad went through. My dad still had to do the milking chores twice a day; my mom still had the other kids to take care of. When they came to my bedside, it must have been like sitting beside a dead boy. Their ten-year-old blond son, his glasses off, his eyes closed. I was unconscious for seven days and seven nights—almost biblical, like Jonah in the belly of the whale, except the whale was in my belly. Then the antibiotics worked.

When I next awakened, it was Sunday morning, but a week later. Surviving this ordeal always seemed a kind of miracle. Dr. Black and Dr. Ceplecha told me to take it easy for the month of December. They gave orders for me to stay in the classroom at school, to not join the other boys and girls for noon-recess play or physical education class for a month. During that time I drew pictures in the fourth-grade room with Mrs. Albrecht. I learned to draw pretty well in that time and loved drawing ever after. And this, too, was when I began to read virtually every Landmark History book ever written, Ethan

Allen and the Green Mountain Boys, George Washington, *and* Lewis and Clark.

In retrospect, this is probably why Dad went easy on me when it came to farmwork. He had me draw the spots on the registration papers for the Holstein calves because I was good at it and liked it, and he often let me out of driving the tractor because I didn't show interest. What mattered was that I had survived; in Perry Kysor's time, I wouldn't have. In part I think Dad was grateful, and in part I think he knew how deeply one could be shaken by a close brush with death.

Farming Forever

HAYING WITH HORSES, HAY LOADERS, & SLINGS

Gilbert Marthaler, German American

Meire Grove, Stearns County, 1924

In June 1924, Gilbert Marthaler had just turned ten years old, but already he had learned to be cautious on the farm, to keep his eyes open for danger lurking. His caution was born partly of his own experiences but was due also to warnings from his parents and stories they told of things that had happened to others in the vicinity of their farm in Stearns County.[1]

That summer, everyone was paying particular attention to the erratic weather. Black clouds in summer brought storms, but at least once a year in Meire Grove, especially powerful thunderheads would roll in, giving the air an eerie greenish tinge just before the storm broke.[2] Gilbert's father, Simon Marthaler, hurried the whole family into the basement when he knew a bad storm was rising; he had witnessed firsthand the power of lightning bolts. Years earlier, one had struck their windmill, splintering and burning its wooden frame.

Such incidents provoked retellings of the familiar story of Benedict Schmiesing, who died tragically at age twenty-two. The children of Meire Grove, including Gilbert Marthaler, listened with both fascination and fear, when their parents and grandparents recounted what had happened to Schmiesing in his hayfield, four miles southwest of Meire Grove, amid the season of *heu machen* (haymaking).

At noon on July 24, 1899, Benedict Schmiesing halted his two-horse team and mowing machine.

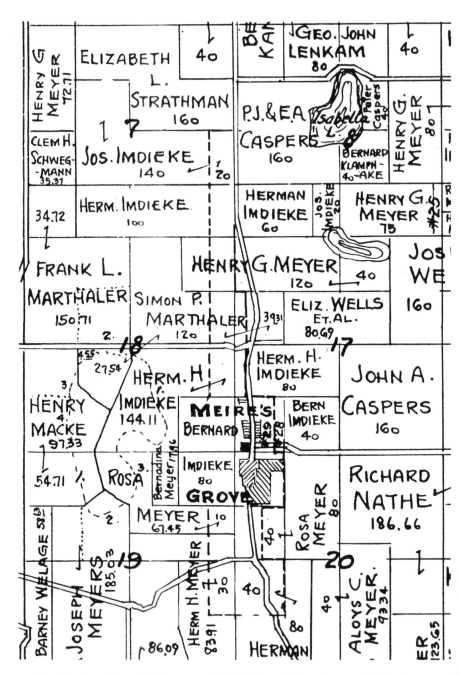

Gilbert Marthaler's farm, still in the name of his father Simon, is shown just north of Meire Grove in this detail of the plat of Grove Township, Stearns County, 1920.
Page 77: Hay sling and carrier as shown in the Louden General Catalog, 1916.

His brothers, Michael and Herman, stopped loading hay onto the hayrack. Their eleven-year-old sister, Mary, had arrived in the hay-field with food and water for their midday meal. After they blessed the food, and as they ate, Mary heard a rumbling in the distance. They looked up and saw a single dark cloud moving on the horizon. Mary became afraid, and said so. As she turned to go home, Benedict told her: "Our Lord can get you at home just as well as out here if He wants you."[3]

When they finished eating, Benedict and his brothers resumed work—Michael and Herman pitching hay onto a stack while Bene-dict mowed. Mary headed home. Halfway there, she ducked under a haystack, hoping to avoid getting wet from the imminent rainfall. She watched as the lone dark cloud swept closer. Benedict, sitting on the iron seat of the mowing machine, pulled on the reins to make a turn no more than fifteen feet from where his brothers stood. Just as they were wondering how wet they might get in the coming shower, a thunderbolt erupted that, they later remembered, took the form of a ball of raging fire—a phenomenon known as "ball lightning." The blast struck Benedict full on the back of his head, "stripping his cloth-

ing to his waist and tearing them into rib-bons," one newspaper reported. He was killed instantly. One of the horses pulling the mower also was killed. The other horse balked but remained in her traces, unhurt.

Herman and Michael stood numbly in utter shock and disbelief. Before they could grasp the violence of what had just oc-curred, another thunderclap rang out, and they crouched, watching lightning hit and set ablaze a haystack across the meadow, in a field belonging to their neighbor Henry Haverkamp.

*Gilbert Marthaler
on his communion day,
about 1924.*

Sudden, accidental deaths always came as a shock in this close-knit farming community, but given the dangers inherent in farming, both natural and man-made, they were not altogether unexpected. There was a measure of equanimity accompanying such seemingly untimely ends, in keeping with the fundamental Catholic belief, as expressed in Ecclesiastes, that there was "a time to be born, and a time to die," and that a person's allotment of time on earth was established before birth. They also became object lessons for future generations to never become careless or proud.

For Gilbert Marthaler, 1924 was a year in which many such lessons were learned. Although the farmwork fell into its regular seasonal patterns, tragedy and loss occurred within a larger pattern of accidents caused by human error when working with machinery. The ten-year-old found a whole world of farmstead fears and joys intermingled on his father's farm just a half mile north of the German-American village of Meire Grove.

The village got its name from the immigrant brothers Herman and Henry Meyer, German Catholics from the province of Oldenburg. They arrived in Stearns County in June 1858 and laid claim to 160 acres apiece of prime property, complete with stands of hardwood trees, easily available water from the nearby Sauk River, and plenty of open prairie land. The Meyer brothers were among those who came to Minnesota in response to the call of Father Franz Pierz, a Catholic missionary priest, for German-speaking settlers to create a Catholic colony in Stearns County. Pierz wrote letters to newspapers in Cincinnati and wrote pamphlets that circulated in Germany, inviting German Catholics to the St. Cloud area. The county quickly became a magnet, and eventually it became the county with the greatest percentage of Catholics in Minnesota.

The church became the center of faith and community in the agricultural hamlet that became known as Meyer's Grove or, officially, Meire Grove. In 1864, the new German settlers helped build a log church on Henry's property. Just seven years later, a total of fifty families had settled near the Meyers, necessitating the construction of a larger wood-frame structure, named the St. John the Baptist Church. In that same year, the St. Paul, Minneapolis, and Manitoba Railway connected nearby Melrose (just six miles away) to the Twin Cities, bringing even more Germans to the region. The congregation doubled during the 1870s and built a new, larger brick church in 1885 on land donated by Henry Imdieke from his farmstead. Even though the Meire Grove settlers were not all from the same province in Germany, they were united by their common Roman Catholic faith.

The Germans of Meire Grove were also bound together by their language. The children received elementary schooling in the German language, so they could read the Bible and memorize the lessons of the Catechism. They heard German in the church observations of Christmas, Lent, and Easter, at weddings, funerals, and First Communions. And they spoke it at home, where the children were also steeped in German tradition. Most of the immigrants were farmers and their children would become farmers. Farming was at the heart of Meire Grove, and the location of the Meyer farm on Meire Grove's south edge and the Imdieke farm on its north edge made it resemble an Old World agricultural village where the farmers lived in the hamlet and worked in their adjacent fields.

Cows became important for the vitality of the tiny community. In the first years, most of the farmers near Meire Grove grew wheat as their main crop, both for flour and for cash income. The 1880s brought overproduction of wheat as new lands on the Great Plains opened to wheat raising, thus driving down prices. But the same decade also brought better railway access and refrigerated freight cars, which opened new markets for butter and dairy products. Dairy farming became the choice of Stearns County's German-American farmers, and they gradually increased their numbers of dairy cows in the period after 1885. Meire Grove's farmers organized a local cooperative creamery in 1897, turning the raw cream into golden butter and delivering it by wagon to Melrose, for transport by rail to distant markets. They built bigger barns and milked more cows. By the 1920s, Stearns County became the number-two dairy-farming county in Minnesota, behind Otter Tail County, which was less suited to row-crop agriculture because of its hilly landscape. In 1924 cows actually outnumbered people in Stearns County, 65,702 to 55, 741. The average dairy farm had between a dozen and twenty cows. With so many cows and so much cream, it is easy to see why, before long, the creamery's importance in the community became second only to that of the church.

Gilbert's grandfather, Joseph John Marthaler, emigrated from Leimersheim in Bavaria in 1846, when he was twenty-three years old. Joseph married Anna Knoeter, also a German immigrant, in New Orleans on January 4, 1849. Ten years later, after having lived in New Orleans then in Illinois, the couple journeyed to Stearns County. Joseph farmed first in the vicinity of Richmond, sticking it out through the Dakota Conflict of 1862—a time when many people fled the area, never to return—then moved to Meire Grove in 1868.

Funeral card for
Joseph Marthaler,
1823-1914.

He purchased 320 acres of land from another German who wanted to move on to other frontiers.[4]

Marthaler's family grew large, finally numbering seven boys and four girls, and there was plenty of work for everybody. Although Simon was the youngest, it was he who ultimately took over the family farm, acquiring the land in 1900. This was one year after he married Rose Schwegman in the village church.[5] The couple had nine children, six sons and three daughters. Gilbert, born in 1914, was the youngest boy.[6]

All the children began doing farm chores at about age seven or eight. The boys started driving teams of horses when they were judged old enough to handle them. Gilbert was first given the reins at age seven, during haying season. Horses were an integral part of farm life, and so farmchildren routinely developed the skills needed to work with them. Gilbert had started out by helping feed oats to the horses before he was seven; he watched how his father and brothers worked with the team, and he followed their example.[7]

The Marthaler family made hay in 1924 with a full line of machines powered by horses: a mower, a side-delivery rake, and a hay loader. Fifty-two-year-old Simon had a deep loyalty for International Harvester equipment, buying his machines from the local dealer in Greenwald, the German community located two miles directly south. Simon needed six horses (three teams) to get all of his farmwork done: Fly and Ginny, both mares; Dick and Prince, both geldings; and Jim and Warny, also geldings. All were crossbreeds, part Percheron or Belgian workhorses bred to smaller saddle horses. Though not as physically imposing as purebred workhorses, they combined workhorse strength with the quick feet of saddle horses. And, besides, they ate less than the large thoroughbreds.

Even though Simon had purchased a small Fordson tractor in 1921, he did not use it for haying because all of his equipment was rigged and designed to be pulled by draft horses. A tractor moved faster than horses and would wreck the horse-drawn equipment. He used the tractor for powering stationary machinery that ran off a belt and pulley, for jobs like filling the silo with corn silage, for grinding feed with a hammer mill, even for sawing wood. Most farmers believed that horses provided cheaper power because they ate the hay, oats, and corn that were raised on the farm.[8]

Marthaler harnessed Fly and Ginny to his mowing machine, for they had lighter feet and a little extra spunk that made them pull a bit faster than the other horses. Simon drove the team for mowing when other work was not pressing. When other responsibilities called him away, he assigned either of his twin sons, Lawrence or Ervin, now sixteen and old enough to handle the team, to run the mower for him. Operating a horse-drawn mower was straightforward work, but there were a few tricks involved in doing it well. The key point was to keep the many triangular teeth on the mower's long sickle-bar as sharp as possible. Sharp blades cut the grass easily and reduced the amount of effort that the horses had to expend, just as a well-honed scythe would ease the labor of the person who wielded it. Marthaler sharpened the mower sickle twice a day when he was cutting hay, usually as the first task in the morning and then at the noon break. The process involved pulling the sickle bar out of the mower and sharpening it with a grindstone. Most farmers kept the grindstone under a tree and performed this chore in its shade.[9]

Young Gilbert supplied the muscle power, turning the grindstone while his father held the sickle against it. This was not a pedal-powered grindstone, as some farmers had, but a hand-cranked one. The amount of strength he needed to turn the heavy wheel depended upon how hard his father pushed the sickle on the surface of the grindstone. Above the stone hung a pail of water with a nail hole in it; the water trickled down onto the grindstone, keeping it wet and thus making the work easier.

Simon was careful to cut hay at the right time of day. His June-grass (bluegrass) was hard to cut well when it was wet for it would "gum up" the mower. So Simon waited until the morning dew had evaporated. The mower needed to be kept going at a fairly brisk pace, for if the driver had the horses go too slowly, the mower would plug up with grass. When this happened, he backed up the horses and took a little run at it to clear out the sickle and keep the blade cutting better.[10]

Another source of problems for mechanical mowers was the dirt in aboveground pocket gopher mounds. The soil got caught in the sickle blade and the blade had to be cleaned by lifting it over the mound, backing up and then moving the sickle back and forth to clear out the dirt. The dirt also dulled the teeth of the blade. Simon Marthaler paid his boys ten cents for each pocket gopher that they trapped. Gilbert earned many a shiny silver dime in this way. He also caught striped gophers by making a string noose for encircling the hole. When the curious gopher popped its head up for a peek, Gilbert jerked the noose around its neck.

The mower was a danger to cats and dogs. No pet was ever cut by the Marthaler's machine, but occasionally the mower killed nesting prairie chickens.

Simon cut the hay efficiently by making his horse team move at a fast walk, going a little faster or slower according to the kind of plants he had to cut. He made hay from grass for his horses, as did farmers from earlier days, but he also made hay from clover or alfalfa for his dairy herd of twenty-five Holstein cows. At two tons of hay per cow, Simon needed to put up over fifty tons of hay as winter feed. He got a massive crop of hay by planting alfalfa seed, which, although it cost far more than grass seed, contained more protein and gave a greater total tonnage per acre. But alfalfa stems were thicker than grass stems and the alfalfa tended to intertwine and get tangled while growing, making it harder to cut.[11]

Even though Marthaler had a mowing machine, he continued to keep a scythe hung on a nail on the machine-shed wall. Scythes had largely gone out of use by this time and only Simon among the Marthalers could swing it well, for he had long experience with it. He used it mainly for cutting weeds around the farm, but he also used it to cut grass in tight places—such as next to fence lines—that a machine could not reach. The boys learned only the basics of scythe handling for cutting weeds, but Gilbert "never could handle that thing" very well because he was a shorter man.

After the grass or alfalfa had been cut, Gilbert or Ervin hitched one of the horse teams to the family's International Harvester side-delivery rake, which gathered the hay into windrows for drying. The side-delivery rake, which had been introduced in the 1890s and was in common use by 1910, was a valued improvement on its predecessor, the dump rake, which collected hay with curved tines and dumped it into uneven piles, which could be made into haycocks or left in rather uneven rows. The dump rake dragged the hay over the ground, losing some of the nutritious leaves in the process. Such a rake worked acceptably well when a farmer hoisted the hay onto the hayrack with a pitchfork, but it was unsatisfactory for producing even windrows.[12]

By welcome contrast, the side-delivery rake produced straight, regular windrows. It was intended for use in tandem with mechanical hay loaders, which picked up windrows best when they were long and even. A side-delivery rake did not so much drag the hay against the ground, instead it lifted, rolled, and fluffed up the hay so that it could dry very quickly. A team of two horses pulled the side-delivery rake; it had a ground-drive system that transferred power from the wheels to the machinery. The rake was ten feet wide, enabling it to sweep together two widths of hay cut by a mower into a single windrow at once. Its seventy-plus single-pronged curved teeth lightly whisked the hay to one side, keeping the precious leaves mostly intact. The hay loader then picked up the hay from the rows.[13]

The Marthalers' old dump rake, like the one shown at left, created uneven piles of hay that were difficult to rake and would clog the hay loader, but the new side-delivery rake, similar to the one at right, produced even, manageable windrows.

During the haymaking season of 1924, Gilbert's job was to drive the horses that pulled the hay loader, a machine that picked up the hay from the windrows and put it on the hayrack. His brother Ervin handled the hay once it got onto the hayrack, lifting and pushing it from the loader to the front and corners of the rack with a three-tined metal pitchfork with a hardwood handle.

Marthaler's International Harvester hay loader, purchased new in 1919, was considered a major advance in haymaking, for it meant that workers would no longer have to make or hoist haycocks by hand onto the hayrack. Although practical hay loaders were being produced as early as the 1890s, their price put them beyond most farmers' reach. And newly arrived, cash-strapped immigrant farmers of course had to rely upon the collective muscles of the entire family to get the hay from the field into the hayloft. The hay loader came into wide use after 1900 as farmers came to appreciate its labor-saving capabilities and equipment manufacturers produced them on a large scale. Prices came down as a variety of companies, including Sears, Roebuck & Company in its catalogs, sold them competitively. The International Harvester Company made mowers, side-delivery rakes, and loaders and advertised the mechanical marvels as the way to reach "agricultural prosperity."[14]

Marthaler's two-wheeled hay loader operated via a combination of horsepower, well-engineered moving parts, and some muscle power. The horses pulled the hayrack with the hay loader attached at the rear end. The hay loader had two tasks. First, to lift the hay from off the ground by means of a revolving rake; all the rake did was to pick up the hay and deliver it to the conveyor. Second, the conveyor

New Idea hay loader with a full load of hay.

then carried the hay up and over the rear of the hayrack. The conveyor resembled a modern-day escalator, continuously moving the hay upward by means of tines that caught and carried the hay.[15]

The horses, Jim and Warny, were harnessed about three feet apart so that they would straddle the windrow and not step on the hay. They were the oldest and slowest of the workhorses. When it came to choosing horses to pull a hay loader, farmers believed, "the slower, the better"; ideally, the creatures would simply plod along. Gilbert's responsibility was to stand at the front of the hayrack and hold the reins to steer the horses. His main concern was to keep them going at a steady speed, and he often had to restrain them a bit in order to have the machine pick up the hay most efficiently. The horses were

When Russell Michalek was a boy of ten, growing up in a Czechoslovakian-American family on a farm near Blackduck in northern Minnesota, he had one difficulty that could have cost him his life. While loading hay onto the hayrack—his team of horses pulling the rack, a mechanical hay-loader behind that—Russell felt proud to be old enough to drive the team. He did well guiding the team as they walked astraddle the windrow of hay. As his father, Andrew Michalek, moved the hay from the rear of the hayrack to the front and sides, he looked at Russell with approval. All was going very well until the horses trod upon a nest of yellow-jacket hornets that were hidden in the row of hay. The nasty hornets arose with a fury, stinging the team of horses on the legs and neck, causing them to react in the only way they knew—to run as fast as they could. The boy tried to hold them, to slow them, to halt them by pulling on the reins of terror. The runaway team bolted forward to the other end of the field. So well trained were they by Russell's father that they followed the row straight ahead, never wavering from their duty, but going way too fast. Andrew Michalek held on to the hayrack for dear life all the while, as the hay loader picked up hay and spat it in an ever-increasing flow onto the hayrack, covering the him with a torrent of hay.

The potential disaster ended as abruptly as it began. When the team got to the end of the row, it stopped. Just like that, the emergency became a cautionary tale as father, son, horses and machines alike all escaped damage or injury. Other farmers did not always escape unharmed in like episodes.

Based on conversations with Russell Michalek, Blackduck, MN, February 27, 1993, and Ruth (Needham) Michalek, Blackduck, MN, July 27, 1997; notes in the author's possession.

well trained, and had pulled the hay loader enough times to know how to follow the windrow. Gilbert gave the horses simple commands, in English, mainly "Whoa" and "Giddap." There was no need to use a whip. If they momentarily disobeyed, he might use a "different language, too," employing curses in German.

Ervin's work on the hayrack involved some hard manual labor. He had to distribute the hay evenly to the four corners of the rack.[16] His first duty was to place a hay-sling directly upon the wooden floor of the hayrack and then cover it with hay to a depth of about four feet. A sling consisted of crisscrossed ropes and several wooden slats, in an arrangement resembling a large net or hammock. After Ervin had completely covered the first hay-sling with a layer of hay, he then quickly spread a second sling over the top of that first layer. Next, he moved hay from the hay loader's conveyor to completely cover the second sling with another layer of hay, again to a depth of four feet. Finally, he spread out the third large sling over the top of the accumulated hay and then covered it with a layer of hay as before, and with that the load was complete.[17]

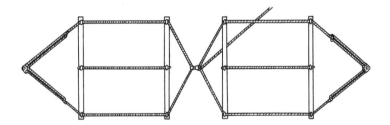

Two popular designs for hay slings in the 1910s.

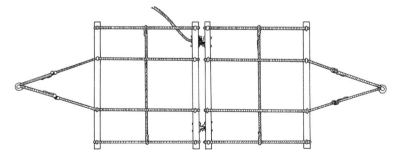

The biggest problem for a hay loader was getting too much hay in the machine at once. If the windrows were too thick and high, the machinery would get clogged and stuck tight with hay and the moving parts became unworkable. Because the wheels provided the power, the wheels would also stop moving, meaning hay had jammed the raking mechanism. If the driver did not stop the horse team the locked-up wheels could break. Once the horses had been stopped, the clogged hay had to be cleared from the machinery with a pitchfork. This problem could be avoided by making sure while raking that the hay in the windrow was not too thick or, after raking, to move some of the excess hay from the thickest spots, carrying it on a pitchfork to spread it out on thinner parts of the windrow.[18]

Even with the hay loader, haying was still hot summer work—for Ervin, especially. But both boys were dressed for haying; they wore loose-fitting short-sleeved shirts, trousers, and straw hats. On hot days they couldn't avoid the blazing sun, but they tried to keep hay from falling down their necks: hay was especially prickly when they were sweaty, and it was hard to get out. They brought a gallon jug of water to the field, placing it out of direct sunlight. If the boys were loading hay at three-thirty or four, one of their sisters, Mayme (age seven) or Edwina (age thirteen), would bring them an afternoon lunch of dried-beef sandwiches—and sometimes cookies—in an empty syrup pail, along with hot coffee.

When the hayrack was filled, Gilbert and Ervin unhooked the hay loader, leaving it in the field, and drove the full load home. Both of them sat atop the stack; and the horses, relieved to head home, set off at a brisk walk. This was enough to create a breeze for those who sat on the load. For farmchildren, the exhilaration of height and the feeling of accomplishment gained from riding homeward high on a hayrack was unmatched by any other part of farmwork.

When their father saw the boys and their load approaching the farmyard, he walked into the barn and climbed up the wooden wall-ladder to the hayloft. On a hot July afternoon, the air up there was oppressive, almost breezeless, still and dusty with its ghosts of the previous year's old hay. It was a sweatbox.

Simon Marthaler had built the dairy barn in 1902 for $500. It was a considerable amount of money but a necessary investment to shelter his livestock. It was large barn, fifty feet by fifty feet, and a full forty feet high. The first floor had space for the workhorses, the cows, and the bull. The upper-story hayloft was roomy enough to store plenty of hay, providing nutrients to the milkcows. The hayloft kept hay far better preserved than outside storage in haystacks. Marthaler, dependent upon mechanical hay-lifting equipment to fill the loft, installed a steel track on the ceiling, which ran the entire length of the new hayloft. On this track was a device called a hay carrier, much praised in its day as a laborsaving marvel. The hay carrier slid on wheels along the steel track, allowing Marthaler to hoist a sling-load of hay and drop it where he chose, along the course of the track, to the hayloft floor.[19]

The boys parked the loaded hayrack at the south end of the barn where its large hay-door was open. They maneuvered it as close to the barn wall as possible, under the triangular hood that projected about five feet from the barn. Gilbert brought the other pair of horses, Fly and Ginny, over near the barn and he and Ervin attached the rope for the hay carrier to the team's singletree and harness. Led

The Marthalers' barn remained unpainted during the Depression when money was too tight to buy paint.

by Gilbert, the horses walked forward and the rope, by means of pulleys, pulled the hay carrier to the end of the track at the peak, so that it was directly over the middle of the hayrack. Ervin then connected the two ends of the uppermost sling, full of hay, to the hook on the carrier. Two metal rings, one on each end of each sling, had to be linked securely to the hay carrier with a *click*. Ervin then gave the signal for Gilbert to move the horses ahead so that the sling-load of hay could be lifted up to the peak of the hay-hood, through the gaping hay-door, and into the hayloft along the track by means of a rope-and-pulley arrangement. Once the horses had pulled the hay carrier to the point where Simon wanted the hay dumped, he hollered "Whoa," and Gilbert halted the team. Ervin, on the hayrack, then pulled on the trip rope to release the trigger of the sling-load of hay, and emitting a heavy *whoosh* it dropped to the hayloft floor with a resounding *whump*. The falling hay created the only air movement in the hayloft.[20]

As Simon tossed the hay into empty areas of the loft, Gilbert turned the team around and returned to the starting point, bringing the hay carrier back outside the hay-door, ready for attaching the second sling-load. Fly and Ginny were so experienced at loading hay into the barn that ten-year-old Gilbert had no problem controlling the team. From the horses' perspective, it probably seemed that they were letting the boy work for *them*.

The Marthalers repeated these actions with the three slings on every load until there was nothing but a little loose hay and some grass seeds or alfalfa leaves left on the hayrack floorboards. Ervin rearranged the bottom sling on the hayrack, hung the other two slings on nails on the front boards of the rack. The brothers were now ready to fetch another load from the hayfield.

In the barn, their father, using a pitchfork, lifted, threw, or pushed the hay into the corners of the haymow and to the sides, keeping it level and ready for the next loads. His work took the most effort and expended the most sweat and energy of all haymaking tasks. Some families put children or the least-skilled hired man in the haymow, because it was so hot in the hayloft. However, because Simon

Marthaler wanted the hay packed tightly and safely he performed that chore himself. If the haying got rushed and he had to put up loose hay that was a bit wet, he would put salt on it to make sure it dried as it should. Marthaler also kept salt at hand on the hayrack in the field in case he needed to spread it on wet hay there.

Marthaler took these precautions because he was aware of how easily spontaneous-combustion fires could erupt in haylofts. For instance, such a fire had broken out in his neighbor John Imdieke's barn not long before, burning it to the ground. Imdieke was unsure how the fire started, but he figured that either hay had gotten stuck in one of the pulleys of the carrier and had gotten too hot, or some hay had not dried sufficiently in the windrows and had heated up after being stored in the loft.

Another precaution taken by Marthaler—like some other farmers—was to forbid his children from playing in the hayloft. The hay chutes were kept closed unless the children went up expressly to throw down hay for the animals. Simon Marthaler was a strict father and what he said was considered law in that German-American household, just as it had been for his father before him. As Gilbert told me, "He told us what to do, and he meant it."

Each season, Marthaler began storing hay at the north end of the hayloft, filling it toward the front as the season progressed. Since alfalfa hay was the first variety ready for cutting—during the first week of June—it was the first to be stored. Packing hay in the loft on a July day always stirred extreme thirst. When they came down the ladder, hayloft workers quenched that thirst with a long draft of cold water from a galvanized dipper, drawn either from the pump or the well. They would splash their hot faces to cool down.

The family made enough hay in 1924 to fill the barn loft, but they needed to pile still more of it on outside haystacks in order have enough to sustain their livestock through the winter. Since pulling hay together for these stacks required less precision than did making windrows, the Marthalers could here rely on the old dump rake.[21] Gilbert and his father pitched the hay from these piles onto the stack, with an older brother arranging the hay at the top. Simon, who stood six feet tall and weighed more than two hundred pounds, had great strength for pitching up hay.

Marthaler did not buy factory-made canvas covers for his haystacks. These were like small dome-shaped tents that protected the large stacks from rain and from wind. Instead he protected his haystacks from blowing winds by strapping the top with ropes tied to wooden fence-posts on either side. With several ropes placed about four feet apart, the wind couldn't blow the hay away. In addition, he told his children in no uncertain terms to stay *off* the stacks. Their feet could easily have made indentations that would let rain penetrate the stack, which would ruin some of the hay. In late fall or during the winter the Marthalers loaded these stacks onto hayracks—mounted on a sleigh or on wheels, depending on the amount of snow on the ground—and hauled them to the barn, getting the hay inside by means of the sling and hay carrier previously described.

Simon Marthaler needed plenty of hay to feed his high-quality Holstein cows. He began purchasing the black-and-white animals around 1919. By 1924 he had just over two dozen of them. Before 1919 his herd had consisted of "red and different color" common "scrub" cows whose milk production was quite poor. The Holsteins produced larger quantities of milk, and during the 1920s the family milked them all by hand. They kept one bull for breeding the cows— and one was enough. Once when Lawrence and Ervin, the twins, were boys they had a frightening run-in with the bull. Told by their father to spread salt on thistles in the pasture, they took their new dog with them and did the work quickly. Knowing the bull was with the cows in that pasture on that day, they each kept an eye on it. All of a sudden, the creature came charging at them. One brother made it through the fence, but the other wasn't fast enough. The boys hollered, "Sic 'em," and the dog went after the bull, distracting it long enough for the second boy to get away. "If it wouldn't have been for the dog," Gilbert recalled, "my brother wouldn't be living any-more."[22] When the twins explained their close call to their father, Simon promptly went out to the pasture with a shotgun. If the bull had come after him, "he would have let him have it," turned the beast into roast beef, and then gone shopping for a better bull. But the bull backed off, and Simon spared its life.[23]

Others in the vicinity that year were not so fortunate. In February Emil Sundflot, a Scandinavian who farmed four miles south of Brooten, died after being gored by his "enraged bull." By the time his brother Arthur had gotten hold of a pitchfork and chased the beast away, Emil's head and face had been "mangled into an almost unrecognizable shape."[24] Over in Sauk Centre a vicious bull killed a visiting Villard man, twenty-six-year-old Percy Tate.[25] And just west of Melrose, Henry Kruse escaped death when his son managed to drive away the bull which had thrown Henry down and stepped on his upper body, breaking a rib and his right collarbone.[26] Such occurrences were commonplace up to the 1950s, when artificial insemination was introduced and a farmer's need to have bulls on hand almost disappeared.

Back in 1979, I heard that Knud Basballe Sr. had been killed loading steers. Knud was a Danish immigrant who came to join the Danish farming community around Morgan quite late in Danish immigration time—1928. He and his wife Mildred, born a Danish American, raised a family on their farm just three miles from the Hoffbeck place. Almost everyone pronounced their last name as "baseball," but it was supposed to be "bahs-ball-uh," and the Basballe family accepted "baseball" as an appropriate Americanization; in fact, how much more American could a family get? His son Carl and I were on the same 4-H softball team that won the Redwood County 4-H championship in 1967, and we were also teammates on the Bethany Lutheran Church League softball team. Knud had played on the team when he was younger.

Knud died when he was seventy-three years old, helping his son Knud Jr. sort cattle for shipping to market. It happened that two steers pushed through a swinging cowyard gate that he was holding, knocking him to the concrete. He sustained fatal head injuries. When I heard about his untimely passing all I could do was to cast my eyes down and sympathize with Carl—I then lived far from Morgan—for his father had passed away in a farm accident just as my father had just eleven years earlier. When children of the farm grow up and move away, they come to understand better how dangerous a farmstead can be.

Other hardships, minor and major, took place in the Meire Grove area in 1924. A July tornado whirled through the area, missing the Marthaler farm by one mile, but it hit one of their large haystacks in the field, demolishing it and scattering the hay far and wide. The funnel cloud touched down and hit Joseph Meyer's round barn, moving it off its foundation but causing no other damage on the farmstead. The same tornado smashed into the barn of Frank Faber, near Greenwald, tearing it off its foundation and dashing it to pieces.[27] Although no one was struck by lightning that summer or fall, several area buildings were hit. One bolt blasted Fred Botz's barn near Sauk Centre, causing a fire that consumed the barn and the twelve tons of alfalfa hay inside it. Botz had managed to untie his horses and get them out safely. Lightning from a late September rainstorm struck the barn at the Theresia Lauback farm in Melrose, and it, too, burned down.[28]

As ever, there also were farm accidents that year, at least one of them fatal. Twenty-year-old Edward Meyer of Meire Grove stepped on a nail, which penetrated his foot. Although he was wearing shoes, they offered little protection from infection. By evening, the foot had become badly swollen. Soon blood poisoning set in, and within two weeks he was dead. Meyer's funeral was "heavily attended," and he was buried in the cemetery behind the Church of St. John the Baptist.[29] Sorrow also came to the Marthalers with the death of Gilbert's maternal grandmother. Caroline Schwegman, who was born in Germany and immigrated to Minnesota in 1885, died at age sixty-nine from stomach cancer. Grandma Schwegman lived a mile and a half from the Marthaler farm. She, too, was buried in the church cemetery.[30]

Simon Marthaler took time to attend funerals and occasional public events in Meire Grove, such as the large-scale annual Fourth of July celebrations. But for him and other dairy farmers, life revolved around work all year long. Caring for milkcows demanded constant attention. Thus he never went hunting or fishing, much less took a vacation. Sundays, however, were kept holy. Accordingly, haying and other farmwork took place during the weekdays, but only necessary

Tornadoes could demolish in an instant even the sturdiest, well-built barn. A tornado hit the Charles Edwards farm near Tracy in 1935 and "completely wiped out" his farm buildings. His place looked "like it had been shelled by powerful guns, with nothing remaining of the barn but the concrete foundation and the floor."

Back in 1883, a tornado hit the Rochester area, causing frightful destruction by hurling, twisting, and leveling everything before it. The funnel cloud came out of clouds that at "first assumed a greenish tint; then copper, then bronze, with whitish angry edges." The "inverted whirling cone" entered the city with the "speed of a cannon ball," lifting "trees, stones, animals and debris" within its "curling mass." It killed thirty-two people in Rochester and then struck the countryside, touching Ole Anderson Moldes's house and barn near Kasson, leaving both "entirely destroyed." Houses, barns, and machinery were "blown all over the prairie." A tornado like this one could kill chickens and turkeys so violently that "even the feathers were blown off of them."

Straight-line winds, called wind shears, have gained renown for knocking modern jetliners out of the sky. In 1947 these thunderstorm winds crushed a Redwood County barn "to the ground, breaking the roof in half." The "blasting wind" knocked another barn "down flat"; "squeezed in the roof" on the next one in its path; and then tore off the top part of a new barn, smashed it, and twisted the wreckage "into weird shapes on the ground." Farmers in Redwood County counted a total of forty-one barns lost along a thirty-mile swath of destruction. The only reported injury was to Mrs. Herman Koll of Morgan, who got caught up in the "excitement of the storm" and "fell against a door and broke her left arm above the elbow."

These straight-line winds could hit a farmstead more than once, as they did at the Otto Dahmes farm in Redwood County. One year a wind blew the barn down and Dahmes was going to build one again, but, as his son LaVerne later told me: "They had the rafters up, and the storm came down, and blew it all down again."

From "3 Persons Injured, Farm Buildings Razed Sunday in Tornado Near Garvin," Redwood Gazette, June 13, 1935, 1, 12. "The Cyclone At Byron," Fergus Falls Weekly Journal, August 2, 1883, 1. "Descriptions of the Storm," Fergus Falls Weekly Journal, August 30, 1883, 1. "Wind, Rain and Hail Cause Heavy Loss," Redwood Gazette, July 1, 1947, 1.

chores were done on Sunday. Sundays were for church. There the
families participated in the Eucharist (Holy Communion) and com-
muned socially with their German relatives, friends, and neighbors.

But disaster struck here, too, when the large and splendidly ap-
pointed brick Gothic revival Church of St. John the Baptist burned to
the ground after a blizzard, the night of February 13, 1923. The
parishioners immediately set about rebuilding it. Construction took
more than a year. The dedication celebration of the rebuilt church,
which took place in October 1924, was unquestionably the highlight
of the year. The Right Reverend Abbott Alcuin of St. John's Abbey in
Collegeville gave a short address following the solemn High Mass.
After services, the community held a parade, with dinner and supper
provided by the women of the parish. The Meire Grove band played
a concert fitting to the occasion.[31]

Whenever the Marthalers went to church they passed the
Imdieke farm. The farmhouse was only thirty yards from the church,
the barn just half a block north. Prevailing westerly breezes blew the
cowyard odors away from the church, but stiff northerly winds
brought whiffs of the cows right to its front doors.

Every second day, the Marthaler boys brought cream from the
farm to the Meire Grove Cooperative Creamery, just across the street
from the church. In 1924, Stearns County and Minnesota were in the
midst of the era of farm co-ops.[32] Established in 1897, the Meire
Grove Cooperative Dairy Association changed farming practices
around the region.[33] While the pioneers depended upon wheat as a
main crop, the creamery made dairy farming profitable. Rather than
being dependent upon income once a year from a single crop, dairy
farmers received regular monthly proceeds from the sale of their
cream. The Marthalers had helped organize the creamery, and Simon
still served on its board of directors.

Another important factor in the lives of Meire Grove residents in
1924 was Prohibition. The Volstead Act, named for the famous Minnesota Congressman Andrew Volstead, enforced the Eighteenth
Amendment to the U.S. Constitution in 1919. In 1918 voters in
Stearns County had come down heavily against the law: 5,322 cast

their ballots in opposition while only 2,345 voted for the amendment. The vote among the Germans of Meire Grove was 131 votes against Prohibition and only 23 in favor. In the German-based culture of Stearns County, taking one or two drinks was acceptable, but "drunkenness was a sin."[34] The main idea was, "Don't overdo it."[35] Not surprisingly, beer was the alcoholic beverage of choice. But with Prohibition, the nearby Melrose Brewery Company had to shut down, as did the Cold Spring Brewing Company, along with five breweries in St. Cloud.[36]

Despite Prohibition, farmers around Meire Grove made home-brewed beer, and some made moonshine whiskey because it brought good profits. Some people around Meire Grove and nearby Elrosa hid their moonshine "in a haystack out in the fields" or in postholes in the ground.[37] Because so many were involved in making beer and moonshine, it was understood that "everyone looked out for each other" to be sure that local bootleggers were not caught by the authorities.[38] When federal revenue agents came for a raid in the Meire Grove vicinity, the telephone party-line served as an early-warning system. The Melrose people called the Meire Grove people and said, "*Riecht nach Lumpen,*" which meant "smells like rags," and then everyone quickly knew that "the Feds were in the area."[39]

Although Meire Grove was not the most prolific bootlegging town in Stearns County, it was located close to Holdingford and Avon, communities notorious for making illegal alcohol. County citizens enjoyed telling the story of a priest who asked the children in church, "Who makes the sun shine?" A child said, "God." The priest then inquired, "Who makes the moon shine?" Another boy said knowingly, "Well, that's from Avon."[40]

Prohibition had little effect on the Marthaler family; they continued to drink beer at threshing time. Harvesting wheat with massive steam-powered threshing machines was the biggest project of their summer, and neighbors and relatives from six families came to the farm to complete the job as quickly as possible. Working from seven-thirty in the morning to nine at night, the workers earned their food and drink. When the threshing was completed, the family rewarded

Saturday-night barn dances added spice to rural social life in the Upper Midwest for at least two generations of farm families. They understood that if a barn was a big as the local small-town auditorium, why not use it for socializing?

Barn dances were, therefore, occasions for merriment on many Minnesota farms. Near Chisago City, a Swedish-American farmer built a new barn and held a dance in the hayloft as a "sort of a dedication," recalled Vernon Shoquist, who attended the dance as a child. Indeed, most barn dances were in new haylofts, for a new barn called for a celebration almost like a "christening." A family might hold a barn dance as part of an anniversary get-together because they could have it at home and the emptied haymow provided the most space.

Families who were known for sociability held barn dances in the spring when the hay was just about all gone. "They would get the barn cleaned up, then they would get a band," and then invite the neighbors. In Stearns County in the 1920s, the Gill family had an orchestra: the four brothers played for barn dances all around central Minnesota. The bands played the old-time music: polkas, schottisches, and waltzes, with a fiddle, accordion or guitar accompaniment, and maybe a tuba and a guy on drums.

A Bemidji farmer, E. C. Hess, hired the Minneapolis Syncopators orchestra for a 1921 barn dance, held to "celebrate the event" of his new barn's completion. He held it on a Saturday night in July, just before filling the new hayloft with hay for the first time. Another Bemidji-area farmer had a local orchestra play at his barn dance later that same year.

During the 1920s and especially during the Great Depression of the 1930s, a number of farmers with fairly new barns held dances in their haylofts in order to make extra money. John Wickander, who farmed one mile north of Wheaton, sponsored barn dances and charged admission to pay for the cost of building his new barn.

Liquor was an implicit part of barn dances. Either the guests were expected to bring their own or the host provided it. During Prohibition times, the dances promoted a form of public immorality, and some anti-alcohol families (whether for legal or religious reasons) would not go. The opposite opinion generally held sway among Stearns County's Germans as some hosts would have bottles of forbidden alcohol hidden in "almost any nook or cranny" at their barn dances, well within the reach of the guests.

Although barn dances were usually joyful events, the owner of the barn might feel a tinge of worry about his visitors' behavior. A carelessly tossed cigarette or an upset lantern could ruin a perfect night and a perfectly good barn. Common sense prevailed, however, for almost all guests realized full well a barn's worth.

Barn dances offered romance with the music. In the Olivia area, one farmer had held many dances in his hayloft, but regretted what took place there one night in the early thirties. His son went wayward and got a term at Stillwater State Prison for impregnating a young lady—she was under the age of consent—on the night of one of the dances. Local legend had it that the liaison took place in a hay manger.

The barn dance idea transferred to radio in the 1920s when station WSM in Nashville, Tennessee, began a new program in 1925, *The WSM Barn Dance*, which featured the music of the mountains. Appropriately, the name of its young announcer was George D. Hay. Chicago's WLS radio copied the idea with its *WLS Barn Dance*, featuring a range of music—including Olaf Sorenson, who sang songs in Swedish both on the air and with a traveling troupe that played in Midwestern towns in the 1930s.

Interview with Edwin Loehr, Spring Hill, Minn., by Rosie Olmschenk, August 9, 1978 (Stearns County Historical Society, St. Cloud, MN), tape #1105, transcript p. 9. Letter from Vernon Shoquist, Chisago City, MN, 1996, to the author, 1.

Interview with Herbert Feddema, January 30, 1979, St. Cloud, MN, with Mildred Dumonceaux, Stearns County Historical Society, St. Cloud, MN, tape #1354, transcript p. 15.

"Big Dance in New Barn," Bemidji Sentinel, July 22, 1921, 1.

"To Give Barn Dance Next Saturday Night," Bemidji Daily Pioneer, October 5, 1921, 6.

Letter from Gertrude Johnson, Hammond, Wisconsin, to the author, November 7, 1996, 2, 3. Letter from Vernon Shoquist, Chisago City, MN, 1996, to the author, 1. Wickander in letter from Edith (Wickander) McMillan, Wheaton, MN, November 2, 1996, to the author, 1.

Interview with Ed Beumer, St. Cloud, MN, by Carol Oman, August 7, 1978, transcript pp. 8, 10, Stearns County Historical Society, St. Cloud, MN.

Sean Pratt, "Barn Buff," Western People (Saskatoon), May 2, 1996, 5.

Letter from Laurance Stadther, Olivia, Minn., to the author, November 23, 1996, 2. "WLS Barn Dance Artists at State Saturday and Sunday," Ward County Independent (Minot, ND), June 20, 1935, 2; Grand Ole Opry souvenir program, 1996.

the crew with bottles of home brew. However, only those who were old enough to put in a full day's work were entitled to beer. Gilbert knew that when he turned sixteen he would be able to join in the threshing celebration, because he would then be big enough to pitch bundles alongside his brothers, cousins, and neighbors. Threshers were allowed beer as a beverage, but haymakers had to be content with water.[41]

The social life of the Simon Marthaler family in 1924 revolved around the church and family visits. Simon's siblings and their families all lived near Meire Grove. However, a neighbor might have a barn dance to celebrate the completion of a new barn. The Marthalers had a good time at these, dancing and listening to "old-time music" performed with a "concertina and a drum, or an accordion." In Simon's youth, fiddlers had played at such get-togethers, but they appeared less frequently in the 1920s.

Nineteen twenty-four was a good haying year for the Marthalers. The hay they stored in the hayloft fed the Holsteins and gave energy to the workhorses through the fall plowing and the long winter. The boys fed the cows twice a day, taking alfalfa hay from the far end of the barn for the first feeding and timothy hay from the near end of the barn for the second daily feeding. The horses ate mostly timothy hay.

For young Gilbert, 1924 was a year of several powerful, extraordinary events: his grandmother's death, a church dedication, and a tornado. It also was a year that, like others, was dominated by everyday events such as haying, turning a grinding wheel, working with horses and machines, threshing, and other humdrum barn and field work. Through it all he experienced the constant reassurance of his secure place in his large family, of fellowship at the Church of St. John the Baptist, and of immersion in the German-based culture of Meire Grove. All of these combined to form his character and his outlook.

Gilbert Marthaler helped his father with farming on the homeplace until he reached adulthood. Then he farmed together with a brother and began renting the home farm from his father. Gilbert married

Marie Schneider, a young German-American woman from the Meire Grove area, in 1946, and afterward made arrangements to buy his father's farm after Simon retired in 1948. The farm passed to Gilbert in 1950 when Simon died, at age seventy-eight, of "old age and diabetes." Although Gilbert was not the oldest male, he inherited the farm—just as his father had—because he was the only one ready and willing to take it over when Simon retired. His two older brothers had already purchased farms in the county. Gilbert had rented the Marthaler homestead prior to his father's death and at last bought the property from his mother, who died in 1969 at eighty-eight.[42]

Gilbert and Marie Marthaler had two sons and two daughters. The oldest son, Allen, graduated from Melrose High School and went on to college. Gary, the other son, graduated from Melrose High School in 1973 and then farmed with his father until he took over the operation in 1978. Gilbert and his brothers and sisters had gone to school only through eighth grade, as was typical for the time in rural areas.

After Gilbert retired, he and Marie moved to Meire Grove. At this writing, they live directly across the street from the Church of St. John the Baptist. Their lives have fallen into a new pattern of breakfast at home, after which Gilbert goes to the DB Bar and Grill to play cards with other men his age. He returns home for noon dinner and remains there the rest of the day.

Gilbert Marthaler,
about twenty-three years old,
with horses Nellie and Tony.

Stearns County continues to be a strong dairy-farming area. It has been the leading milk-producing county in Minnesota in recent decades, surpassing Otter Tail County. Stearns County ranked fourteenth among counties nationwide in value of dairy products sold in 1992. Logically, Stearns County also was the leading county in Minnesota in tons of alfalfa hay harvested that year, and it ranked thirteenth among counties nationally.[43]

Stearns County continues today to hold the most distinctively ethnic communities in Minnesota. In villages like Meire Grove, the church remains the central feature, along with a cemetery, a small grocery store, and at least one tavern. The creamery buildings are generally still standing but are no longer used as such.[44] The Marthaler place also has retained its strong heritage as a dairy farm, being listed as a Minnesota Century Farm, one that has stayed in the family for more than a hundred years. The family has been farming the same land for five generations in Minnesota, and Gilbert Marthaler believes his family will still be farming there five generations from now. Just as the Marthalers had always been farmers in Bavaria, they would continue to be farmers in Minnesota.

Gary Marthaler built a new barn in 1993 with stanchions for forty-four cows. The new barn was old-fashioned in the sense that it has a huge hayloft. Gary still makes square bales and has a bale conveyer among the rafters in the center of the hayloft. He does not stack the bales in the hayloft; he simply allows the bales to fall in any order. Nothing else on the farm, however, could be described as disorderly. Marthaler plants, tills, harvests, and plows his 191 acres of land in the proper season with the help of his wife, Irene (of German extraction and a Melrose High School graduate), and three sons—Eric, Kyle, and Kurt, one of whom will likely take over the farm someday. A visitor to the new barn will understand just how orderly it is by observing the rear ends of Marthaler's cows. Their tails have been cut off so that a cow cannot dip its tail in the manure behind it and whap Marthaler when he kneels beside a cow to milk it.

Gilbert Marthaler still visits the homeplace often, checking to see how the corn is doing or if the alfalfa is ready for cutting. He walks with the aid of a cane, but tries to see everything that is happening. He and his wife own a burial plot in the village graveyard; the tombstone is already in place, their names engraved on it. When it's their time, they will join the other faithful German Americans buried in orderly rows just east of the church in Meire Grove.

The burial plot for Gilbert and Marie Marthaler, flanked by stones for members of the Schmiesing family, in the St. John's Catholic Church cemetery, Meire Grove, 1996.

Vernon Christian Hoffbeck, my uncle Vernie, served in the Army as a B.A.R. man. That's for Browning Automatic Rifle, and in January of 1945 he used the fast-firing but heavy gun in the retaking of the Philippines. Outside of the town of Lapao, the Japanese pinned down Vernie's platoon with machine-gun fire in early February. At ten P.M., the enemy attacked. Sergeant Erickson told Vernie to get his B.A.R. going. Both men knew that was a fatal order, for

Vernie had to stand directly into Japanese fire. But he did as he was ordered and stood up, firing as the enemy tossed hand grenades at him. One exploded directly in front of Vernie, its shrapnel piercing his chest and abdomen.

So many thoughts rushed through his mind, thoughts of his mother begging all three of her boys at war to come back home again, worries that his groans of pain would allow the Japanese to pinpoint his buddies' locations. Just then, another farmboy from southern Minnesota (near Rochester), Sgt. Adolph Pisny, pulled Vernie to safety and to a field hospital for surgery.

Vernon Hoffbeck, wounded in the Phillipines, in 1945.

Several days later, Sgt. Pisny visited Vernie. The sergeant watched him with sympathy, for he knew Vernie was sure to die. Vernie knew it too and asked Pisny to write home to tell the folks that Vernie would not be coming home. Vernie was only twenty-five years old. The sergeant couldn't bear to do it, and the family only got word that Vernie was wounded—and later that his condition was improving.

Remarkably, Vernie recovered after several months flat on his back, thanks to blood plasma and modern medicines. He married Darlene Dahmes that summer; eventually, the couple had seven children. The last-born were twins, in 1962, a girl named Shelly and a boy named Kelly. Shelly was beautiful, like her three sisters; all became homecoming queens at Morgan High School. Kelly was a bright athletic boy, loved by all for his sunny disposition.

Vernie planned for Kelly to take over the farm, but it didn't work out that way. In early September 1978, when Kelly was sixteen and a junior in high school, he was picking up haybales from the road ditch, hoisting them onto a hayrack, when the tractor tipped over on him. He was dead on arrival at Redwood Falls Hospital. Kelly died almost exactly ten years after my dad did. I should have gone to the funeral, but I told myself I was too far away from home to make it there.

But I knew how deeply my uncle's family grieved over Kelly's death, for Vernie had shared the loss of my dad to a farm accident. Vernie had been close to my dad, trading work with him, worshiping in Bethany Lutheran, and playing cards with him. I have learned that when a young adult dies, a family feels the loss especially keenly, wondering what the departed would have accomplished in the prime years of high school and postgraduation, when they started out on their own. Uncle Vernie and Aunt Darlene kept photographs of Kelly on the piano and in almost every room; they printed a poem in the local newspaper on each anniversary of his death.

His older brother Laurie felt that he might have somehow prevented his brother's accident and took over the dairy farm in place of Kelly—twin burdens—and he worked extremely hard to do all the farmwork, trying his best to replace Kelly. Eventually, he even dated Kelly's high school girlfriend, a dazzling young woman who had been immeasurably saddened by Kelly's death. It was not a good idea for him to date his brother's true love. When they later married, it was still not a good idea. Soon after, the girl's mother fell off her horse and struck her head on concrete. She died from the injury. Laurie's wife couldn't handle the grief this time and soon after the couple divorced. It was too hard for anyone to take Kelly's place, it seemed.

Vernie, who died in 1995, lived to be seventy-five years old. In some ways he never got over Kelly's death. The light in his eyes, a last vestige of my dad, dimmed after the accident. His old wound from the war acted up occasionally, and he got treatment from time to time at the Veteran's Hospital in the Twin Cities. He had never talked about his close call in the Philippines but every January and February he "did a lot of thinking" about it and about the man who had brought him to safety, Sgt. Pisny. When Vernie was seventy, he made an effort to contact Pisny, running a missing-persons advertisement in The

Farmer *magazine. Thirteen people responded to the ad, and the two men met again, forty-five years after the sergeant left Vernie's bedside thinking he would never see him again. After that, Vernie could finally speak about his brush with death on the battlefield.*

MISSING PERSONS

I would like to contact someone who knows Sgt. Adolph Pisny, Co. E, 35th Inf. Regiment, 25 Div. during WWII. I'm not sure of the spelling of his name, but he was a farm boy from, I think, the Rochester or Albert Lea, MN, area. I was in this platoon in the January 1945 invasion of Luzon where I was wounded by a hand grenade on Feb. 4. Sgt. Pisny and a medic helped get me to the road, a jeep and a portable field hospital. After the big battle of Lapao, he came to see me. I am sure he didn't think I would get up again. I can still see the sad look in his eyes. I got up, have been thinking about this for 40 years and would like to find out about him. V. C. Hoffbeck.

The Farmer *magazine* ad, July 1989, 45.

Tradition and Change

TRACTOR-POWERED HAYING WITH A HAY BALER
Arthur and Douglas Rongen, Norwegian American
Fertile, Minnesota, Polk County, 1959

I've liked round barns since the first time I saw one at the Shelburne Museum in Vermont. Over a period of years after that, I gathered information about round barns in the Upper Midwest for a short article I eventually published in *North Dakota Horizons* in 1995. While I was working on the project, I happened to mention it to Cal Lee. Cal was a six-foot-five, blond, Norwegian church-basketball teammate of mine when I was still living and teaching in Minot, North Dakota. He said his family had a round barn on their farm near Fertile, Minnesota, and told me to write to his mother, Marlys Rongen Lee, if I wanted to know more about it. Marlys sent me a few articles about their Centennial Farm with its round barn, and I was intrigued. I liked what I saw about the barn, and I was wanting to include a Norwegian family in this book, so I decided to make a trip out to the Rongen farm.

Art Rongen was ninety-two years old when I met him, and it seemed wonderful to me that the grandfather of the family still lived on the homeplace in a trailer just away from the farmhouse. This old-fashioned family seemed like a good example for showing the transition from making loose hay to baling hay, but I worried the year they made that changeover—1959—

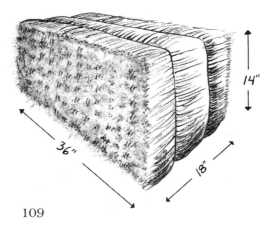

109

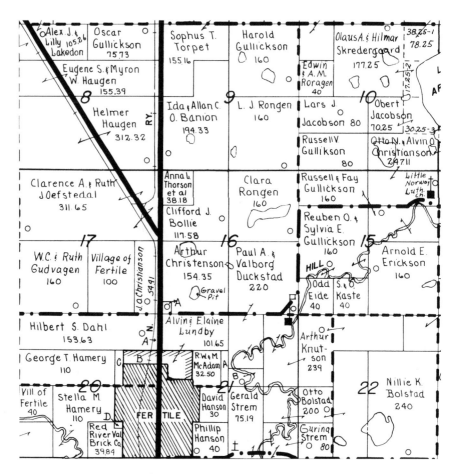

Art Rongen's farm is still listed under the name L. J. (Lars) Rongen in this detail of a plat of Garfield Township, Polk County, 1954.

Page 109: *Standard hay bale.*

was a little late, because so many farmers began to use square bales in the early fifties. But something about the Rongens captivated me. Partly, it must have been the stories everyone told me about Art.

Marlys said that he was infamous for a prank he pulled in the 1940s. He was trying to surprise a pair of newlywed neighbors with a shivaree welcome. "He used dynamite," Marlys said, "and blew out all the windows. It was heard all over the neighborhood." At a Sunday school picnic in the sand hills near Fertile, Art drove the car straight up the steep bank of a bluff with the whole family inside. Even in town, he had such a reputation for speeding and pulling U-turns that he was finally told to stay out. When I interviewed Art, the spark was still there. No sooner had I asked about the old round barn than he was out and climbing right up to the hayloft. Though a man in his nineties, he climbed the ladder without a thought, talking all the way about anything I might need to know about the barn.

His father Johannes Rongen, an immigrant from the Old Country, built the round barn in 1894. It was clearly visible from the road, and those who traveled by may well have wondered why a seemingly sensible man would build a strange structure that contradicted the traditional Norwegian rectangular barn. Most immigrants who settled in this area, known as "Little Norway," wanted to preserve as much

Art Rongen on the 1918 tractor, July 1959.

of their Norwegian identity as they could.[1] The earliest Rongens and their neighbors established a culture that was part American but mostly Norwegian, and they clung to their traditional ways of life for at least three generations.

Evidence of their religious culture—the Little Norway Lutheran Church—stood just half a mile from the Rongen round barn. The congregation had built the church, with its seventy-five-foot-tall steeple, in 1891. Originally, the men sat in pews on the right side of the aisle, and women and children sat on the left side; the church voted to abolish that policy in 1899, but it took another generation before everyone mingled within the church. Worship services were conducted exclusively in Norwegian until the mid-1930s, and occasionally as late as the 1950s.[2] Minutes of church meetings were not written in English until 1934. Change came slowly to the Little Norway Church, just as it did to the families who attended it.

Johannes left his birthplace near Bergen, Norway, in 1875. After spending five years farming near Lake Mills, Iowa, he came by covered wagon to Garfield Township in Polk County with his wife, Agatha, and four children: Synneva, Brynjel, Lars, and Ole.[3] They followed the old oxcart trail into a part of the county not yet accessible by railroad. The Rongen family settled with other Norwegians in a farm community they named Fertile, after the Iowa town that many had recently left. No doubt they hoped the name would be auspicious and that the area's sandy soil would be fertile ground for nourishing heavy wheat crops. Located just twelve miles east of the Red River Valley, which was already being farmed, Fertile was surrounded by rolling hills formed from the sand dunes along the edge of ancient Lake Agassiz.

Johannes Rongen chose a homestead just a mile from the Sand Hill River, near enough for drainage but far enough away to avoid flooding.[4] The Rongen farm had plenty of oak trees, and the combination of trees and sand hills made it more attractive to Johannes than the relentlessly flat land of the Red River Valley, even if the valley had richer, blacker soil. Conservative in all other farming matters, he built his singular round barn fourteen years after settling near

Fertile. Some said he made it round so that harsh winds from the Dakotas would blow around the barn without knocking it down. Others believed that a tornado wouldn't be able to suck it off its foundation. Most likely, Johannes thought a round barn it would be more economical and efficient than a rectangular one. Built to shelter seventeen dairy cows and four workhorses, the barn was "nice for feeding" the livestock because all the animals faced toward the center and the Rongen boys could drop the hay down from the hayloft into the middle of the barn.[5]

To avoid debt, Johannes built his farm structures one at a time.[6] Two years after erecting the distinctive barn, he built a large farmhouse to replace the original log cabin. Apparently, he was heeding an old Norwegian saying: "A barn can help build a house, but not the other way around." Johannes Rongen farmed the land with his son Lars, who took over completely in 1918, three years before his father died. Lars then farmed until *his* son Art inherited the operation. The two generations lived together in the large farmhouse Johannes had built. It was expected that the eldest Rongens would live on the homeplace with the son who inherited the farmstead, just as it was

expected that Johannes's five children would marry Norwegian Americans—which they did.

The Rongens had always kept a few dairy cows for milking, but they expanded their dairy herd after the Fertile Creamery Association was formed in 1901. Lars served on the board of directors of the new organization, a cooperative that made the business of

Art's Dad, Lars Rongen, at the farm near Fertile, 1950.

dairying more lucrative. The Rongens were typically cautious about spending money to modernize their dairy operation. In fact, Lars waited until 1919 to construct a wooden-stave silo for corn-silage feed for the cows.[8] Corn silage supplemented the loose hay that the Rongens harvested—using horses, a mechanical mower, slings, a hayrack, and a hay-carrier track mounted on the hayloft's ceiling— and then stored in their large hayloft.

The Rongens took good care of their old machinery and waited to see if late-model equipment would be more cost-effective before they purchased it. While Lars's son Art farmed with his elderly father, their haying operation continued to be powered by horses. Gradually, however, the family did buy improved tractors: a Titan for plowing and to power the threshing machine in 1920, an F-12 International in 1937, and an "H" model International Farmall in 1950. By the time Art assumed full operation of the homeplace, after Lars died in 1959, he had completed the transition to tractor-powered farming.[10]

During the 1940s and early 1950s, all five of Art Rongen's children helped gather in the hay, using hayracks, pitchforks, and muscle power. Daughters Marlys and Judy worked with their brothers Arvid, Charles, and Doug in the field and in the barn, but as soon as each girl turned fifteen, she received her driver's license and worked as a waitress at Ole's Café in Fertile.[11]

For many years, the Rongens' haymaking routine was unwavering. Grandpa Lars or Art cut alfalfa with the horse-drawn mowing machine during the day. In the evening, the children and the adults, sometimes including Art's wife Hilma, would cock the hay, which had been raked with the horse-drawn dump rake. After the hay dried, they hauled it to the round barn. The Rongens also got wild hay from the 160 acres of prairie land they owned seven miles northwest of their farm; they put this natural prairie grass into haystacks for later transport to the loft in the barn.

Getting hay from "the prairie," as the Rongens called it, was a memorable family tradition. The three oldest children—Arvid, Marlys, and Charles—ventured out there for a week with their father as soon as each was old enough to handle the work, at about age nine or ten. Grandpa Lars stayed home to do the milking and the other chores, while Hilma tended the house and the younger children. Art led the expedition, taking two teams of horses, each of which pulled a hayrack. Arvid and Marlys took turns driving one team—King and Prince—while Art led the way, driving Birdie and Maude because they "were harder to handle."[12]

For Marlys, who went to the prairie for haying during two summers, the adventure became her fondest memory of the farm. Between the ages eight and fifteen, she performed other chores as well, including taking lunches—sandwiches, coffee with cream in a quart jar wrapped in layers of paper to keep it warm, cookies, and doughnuts—to her father out in the field. She also fed the chickens, gathered eggs, and slopped the hogs with buttermilk and ground oats or leftovers from the house.[13] Driving horses was commonplace for young Marlys and her brothers; she especially liked King and Prince, both geldings, because they were so gentle. The children were not allowed to *ride* the workhorses, however, because, as Lars said, "They worked hard enough without hauling you children around."[14]

A week of haying on the prairie began early Monday morning. Hilma would already have prepared and packed enough food to satisfy Art and his young crew for a week. In addition to homemade

Maude and Birdie outside the round barn on the Rongen farm, about 1950, before the equipment was tractor driven.

doughnuts, cookies, sauces, and bread, she sent along milk, eggs, bacon and side pork, beans, cheese, meat, and such staples as butter, salt, and pepper, plus all the necessary dishes, pots, and pans. The food was packed in five- or eight-gallon milk pails and cream cans, which were kept cool in a well on the prairie.[15]

The family worked together from morning until night, mowing, raking, and stacking the wild hay. Each had a job to do, either on the stack, or using a bull rake hitched to two horses to push the hay together into a huge pile, or forking the hay. Art usually mowed and, as the children lifted the hay to him with pitchforks, formed the haystacks. The girls were never allowed to make the stacks and so never learned how to perform that chore.

In the mornings before they started work, Art boiled eggs right in the coffeepot while the coffee brewed. He also fried eggs and toasted bread on the stovetop grill. Human and horses alike welcomed lunch and supper breaks. Each evening the children fed, watered, and brushed down the horses for the night in a little barn on the prairie property. And after supper they wearily tucked themselves into the oats bin, right across from the horses, for a sound sleep. They remained in their workclothes, but the sheets and blankets they had brought made their unusual bed quite comfortable.

At the end of the annual week on the prairie, dozens of large haystacks stood in the field. Before heading home, the Rongens always loaded up their two hayracks. Then, again with Art driving one team, the children the other, they made their way back toward Fertile, going slowly down the hills so they would not overwork the horses or lose any of the hay. Halfway home they stopped at Score Hill to drink some of the cold, crystalline spring water.

Art hauled the rest of the haystacks from the prairie in the winter, when he could use sleighs with hayracks mounted on them. He took two sleighs with two teams of horses, one team tied behind the first sleigh and pulling the second sleigh. He would pitch hay onto the first sleigh in the forenoon, eat dinner at noon, and pitch the second load in the afternoon. He would return home in time for supper.[16]

Getting the hay into the hayloft also was a family undertaking.

Sometimes Marlys drove the team, which lifted the hay on slings into the loft by means of the hay carrier. Arvid would connect the sling ropes to the carrier and tell the driver of the team when to hoist the hay up by means of a rope-and-pulley arrangement. Art, who was up in the loft, had the hardest and hottest task—carrying the hay to the outer edges of the round structure. When alfalfa cultivation became popular in the 1920s, the Rongens always put alfalfa hay on the south side of the loft and wild prairie hay on the north side. The wild hay was easier to throw down the hayloft hole because it "didn't hang together as tight" as did the alfalfa hay.[17]

The family considered the hayloft a "sacred place" because it stored food for their cows and horses. Grandpa Lars continually instructed the children to respect hay as the sustenance of the livestock, asking them, "How would you like to have your food stepped on?" The only areas for play in the loft were those where no hay was stored.[18] Another reason the hayloft was to be treated with respect was to avoid the kind of tragedy that had occurred in 1891. Great-grandfather Johannes Rongen had gone back to Norway to see his mother before she died and brought back with him two nieces and two nephews to live in America. One of the nephews, Lars Fadness, who worked for his uncle Johannes as a hired man, was exceptionally strong and quite acrobatic. At a barn dance near Fertile in the dead of winter in January 1891, twenty-one-year-old Lars attempted to walk on a rope stretched across the hayloft. His stunt ended fatally when he fell to floor, breaking his neck. He was buried next to the family plot in the Little Norway Lutheran Church cemetery. Art and Hilma Rongen never talked to their children about Lars Fadness and his tragic fall, but they impressed upon the young Rongens that the hayloft was potentially dangerous as well as sacred.[19]

Marlys, who was afraid of heights, was never tempted to play in the hayloft; she preferred to have her feet planted firmly on the ground. In the winter, though, she helped pitch hay down from the loft to feed the livestock. Sister Judy, however, enjoyed playing in the loft. Judy never forgot one time when Art let her ride up in the sling when the loft was almost full. Grandpa Lars watched her ascent and

then helped her out of the sling. Charles, always the most daring of
the children, would sometimes jump from the hayloft door and land
outside on a fully loaded hayrack—but only when the adults were not
around.

Work always took precedence over play, however. Marlys loved to
pitch manure, standing barefoot in it while cleaning the calfpens. She
forked the calf manure into the litter carrier, pushed the carrier out
the door, and dumped the contents on the manure pile. But as soon as
she got her waitressing job in town, Marlys was through with haying, and her sister took her place.

Hilma Rongen was a good baker and cook who taught her daughters how to prepare such traditional Norwegian foods as *lefse* and
rømmegrøt (cream and flour pudding); *knekkebrød* (baked flat bread)
and *flatbrød* (grilled flatbread); milk mush, crisp rosettes, *sandbakkels*
and *fattigmann* (two kinds of cookies), *rulle pølse* (beef roll), *spikekjøtt*
(dried beef), and *blodklub* (blood sausage), and, of course, lutefisk. At
Christmas they made the traditional *fruktkake* (fruitcake) and *julekage*
(Christmas cake).[20] Judy, more domestically inclined than Marlys,
won blue championship ribbons for her cooking and sewing projects
at the Polk County Fair.

Marlys Rongen on the Farmall H tractor, July 1954.

Before her children were old enough to help their father, Hilma made haycocks in the cool of the evening and sometimes stacked bundles of wheat during the harvest. But between rearing five children, doing housework and yardwork, sewing for her family, and serving as a volunteer at the Little Norway Church, she had little time or energy left for farmwork.[21] Arvid, the Rongens' oldest boy, and Charles, their second son, worked closely with their father on the farm. But both boys left home to seek their life's work elsewhere after they graduated from Fertile High School—Arvid in 1951 and Charles in 1955. In their absence, Judy continued the prairie-haying tradition with her father until she graduated from high school in 1957. By that time, only young Doug remained at home.

When I made a second visit to the Rongen farm, a summertime visit, Doug took me in his pickup to the "prairie." It was fairly close to the main highway, but no traces of any building remained on the property, no reason to make note of the land if you were merely passing through. Even standing there, it looked to me like nothing but a big, open, grassy field, but to the family it was a special place. I could tell by the way Marlys wrote about it, and I could tell from the way Doug talked and acted as we arrived.

For Doug, the events of 1959 left deep impressions. He was only in his midteens, but he was about to experience a major shift in the ways his tradition-bound family of Norwegian Americans conducted their lives as dairy farmers. Indeed, the 1960s was a decade of change for all farmers throughout the state. He remembered hoping that the acquisition of a brand-new, bright-red McCormick baler was a change for the better. Baling hay with a tractor-pulled machine was bound to be easier than the old-fashioned loose-hay method.

For the first time since the Rongens settled in northwestern Minnesota, they would be working without horses. Their final team, faithful old King and Prince, had been sold for slaughter the year before. Doug's sister Marlys had come home from Mayville, North Dakota, with her husband and their little boy, Cal Ray, to say goodbye to the steadfast animals that had served the family long and well.

They had been a good team and gentle enough to let the children squeeze between them to put on their harnesses. But the era of work-horses was ending, and that of tractor farming was beginning.[22]

The most emotional event of 1959 occurred in July, in the middle of hay baling, when Grandpa Lars Rongen passed away at age eighty-three. Grandpa, the first of Doug's close relatives to die, would be sorely missed. He had regaled his young grandson with stories of the old days, when Chippewa Indians used to come to the Rongen farm asking for water. He also had taught Doug about Nor-wegian culture and customs.[23]

In keeping with family tradition, Lars had continued to live with Doug's family on the farm after he retired in 1956, helping in what-ever way he could. When he was eighty his health began to fail due to a heart condition. Toward the end of his grandfather's life, it was Doug, the youngest of five children, who drove Lars to the farms of nearby relatives for visits. Though Doug was too young to have a li-cense, Grandpa Lars would let him drive anyway—but never faster than thirty miles per hour. Lars Rongen was buried in the sandy loam soil of the graveyard of the Little Norway Lutheran Church.[24]

At the same church, during the same year, Doug Rongen passed from childhood to adulthood when he was confirmed. He had faith-fully attended confirmation classes for two years—every Saturday

Doug Rongen
at age eleven,
1955.

morning from ten o'clock until noon at the church—and he had mastered the answers posed in *Luther's Small Catechism* to the satisfaction of Pastor Thomas Gabrielson. For Doug, as for other Minnesota Lutherans, confirmation was a significant life passage. It meant that, as a full member of the congregation, he could now partake of the bread and wine at Communion. It also meant that the community would start to see him as able to assume manly responsibilities.[25]

Fewer hands at home meant more work than Art and Doug could do alone. In 1959 Art Rongen had little choice but to purchase a baler, made by International Harvester. His nine-year-old tractor was also an International—bright red, with the two wide wheels on each side of the driver's seat, and two narrowly spaced ones in front. It was an H model, a popular first tractor for many Minnesota farmers in the 1950s. In order to make orderly and straight windrows for baling, he also bought a new side-delivery rake. Art Rongen always bought his equipment from Solon Gullickson, a fellow Norwegian who was Fertile's local dealer as well as the mayor. Like most farmers, Art was loyal to one local farm-implement company and one manufacturer. Whereas some farmers favored green John Deere tractors, others, like Art, preferred International's red models.

This 1946 photograph by Myron Hall shows a Case wire-tied baling machine. One man drives the tractor while a man and a boy sit on the back, tending the baling wire. Wire-tied bales could weigh almost double what the twine-tied bales did.

Back in 1937, Art and Lars had bought their first gasoline tractor, an International F-12 model, to do the heavy work of plowing and disking. Driving it took some getting used to—after all, it didn't handle quite like a team of workhorses. Once when Lars was plowing with the F-12, the tractor tipped over because he had not adjusted it for driving over an incline. He managed to get out of the way in time, but he could have been pinned beneath the tractor and severely injured or killed. Despite that mishap, both men liked the tractor and the tractor company, so they continued buying International Harvester machines from Solon Gullickson.[26]

Haymaking with modern equipment pulled by tractors followed a set routine. First, Art mowed the alfalfa in the hay fields around the homeplace with the old mower, converted for pulling by the even older F-12 tractor. After the alfalfa had dried for half a day, Doug drove the F-12 with the side-delivery rake; he raked two swaths together, producing the right-size windrows: not so thick that the cut alfalfa couldn't dry and not so thin that the baler couldn't easily pick them up. After the hay dried in the windrows for about a day and a half, Art would test it for moisture. When it felt dry enough to the touch and, therefore, ready for the baler, Art and Doug knew they had to move fast to make hay before it could be ruined by rain.

Doug, who had driven workhorses since he was nine years old, caught on quickly when, at age fifteen, he drove the new tractor for baling. He did, however, have to learn about the idiosyncrasies of the H. For example, Internationals did not have "live" power takeoff at that time, so when the clutch was pushed in, the tractor engine could not provide power to the pulled equipment. Without so-called live PTO, a tractor was trickier to operate. But Doug soon learned that when he came to a place where the row of raked hay was really thick and heavy, he had to take the tractor out of gear; that way the tractor stayed put long enough so all the hay in that spot could be picked up and cleared through the baler. It was especially difficult to drive the H while baling on a downhill slope: Doug had to pop the machine out of gear while keeping the baler running at the regular baling speed and then ride the brakes to control his speed going downhill, while limiting the amount of hay entering the baler. If he pushed in the clutch pedal, the baler would not pick up any hay, and the tractor, baler, and hayrack would coast—hayless—down the hill. Later models of International tractors were routinely equipped with live PTO, which enabled the operator to push in the clutch while the PTO kept revolving.[27]

While Doug drove the tractor that pulled the baler, which pulled the hayrack wagon, his father stacked the bales on the rack. Art soon learned that his old skills with a pitchfork were of little benefit in moving forty-five-pound bales of hay. To stack a bale that large on the hayrack required a technique: the thrower had to swing back, pendulumlike, and then swing his right leg up to help hoist the bale, which was partially lifted and propelled by its own weight.

Art wore gloves to perform his arduous task of handling bales. The side of the bale that was cut as it moved through the baler felt like the bristles of a wire brush, the hay stems painfully poked his arms and hands, and the twine that secured the bales chafed his finger joints. Art frequently had to replace his yellow cotton work-gloves, stained green from the alfalfa leaves, because they wore out so fast.

Stacking bales on a hayrack was somewhat similar to stacking loose hay on a hayrack. In both cases, the goal was to arrange the hay so that it would not slip off the hayrack when it was being driven from the field to the barn. Although the instruction manual for the baler gave no directions for stacking the so-called square bales on the rack, just about everyone learned to stack them one of two ways.

The International baler generally performed well during the summer of 1959, but Art discovered that the manufacturer's engineers had failed to design a high-quality knotter. The machine was supposed to bind each bale together with two twine strings, using a mechanical knotter to tie and cut the twine in preparation for the next bale. But sometimes it left one or both strings untied. When a bale emerged on the conveyor with one string untied, Art had to stop and retie it. If both knots were improperly tied, he had to yell at Doug to stop the tractor and help him carry the loose bale to the front of the baler and hand-feed it to the pick-up mechanism. When the knot-

Farmers took pride in making large loads of hay, in building mounds of hay on a hayrack or hay wagon that might be a little higher and a little wider than those accomplished in previous attempts. This was true in both the loose-hay period (1850s to 1960s) and in the small-bale period (1945 to the present).

When farmers pitched loose hay from haycocks onto a hayrack, the general goal was to fill the rack efficiently and build the stack as high as possible. The saying of the times was, "Keep the sides full and the middle will take care of itself." This meant that the worker on the hayrack would build up the corners and edges of the hayrack with hay, pack it down with his feet, and then fill up the middle only after the sides were well-formed. It would be foolishness to pile the hay in the middle and then attempt to add to the sloping sides of a mound; the hay would slip right off the rack.

Hayrack design was a part of the folkways of the country, for farmers were concerned with keeping hay on the hayrack but also wanted to limit the barriers to pitching hay up to it. Thus hayracks came to have a tall back and a tall front, with four triangle-shaped extensions along the sides of the rack. This design allowed the hay to be tossed on the hayrack efficiently, and also permitted farmers easy removal of hay from the surface of the rack.

The physical limits to the amount of hay put on a hayrack were fairly obvious: a person on the ground could lift hay only as high as he could extend the pitchfork above his head and throw the hay. A tall, strong hay pitcher could hoist hay quite high.

The other restraint on large loads came from the possibility of hay slipping off the hayrack on the way home from the field. Despite this commonsense knowledge, foolish pride often led some Minnesotans to make a load too high or too wide and part or all of the hay would slide off the rack if it tilted too far while in motion—especially on hillsides. Others succumbed to the vanity of making a load of hay so large that it could not be driven into the oversize doors of a drive-in barn.

Haymakers in the Zumbrota area, who hauled loads of hay through the covered bridge on the Zumbro River, had a similar concern—could they fit their load inside the bridge? Here folk wisdom and folkways had historically prevailed. Covered bridges in Vermont, New Hampshire, and elsewhere were built to a common specification, namely, to make the dimensions "a load of hay wide and high."

When farmers made square bales of hay—which were actually rectangular—they found that the design of the hayrack had to be modified. The boards on the front, where a driver had once stood to hold the reins of the horse team, had to be removed so that the bales could move out of the baler and directly onto the rack. However, the rear design of the hayrack had the same look as that of the old hayrack used to haul hay that had been pitched up from haycocks.

By the time that farmers used balers, starting after the end of World War II, hay wagons began to feature rubber tires, rather than the old-fashioned steel wheels. The rubber tires permitted the rack to move more smoothly over dirt trails through the fields and gravel roads, allowing for the possibility of larger loads of hay.

This photograph of a farmer atop his wagon, about 1910,
shows the limits of the loose-hay method.

One method of stacking bales used a complex system of interlocking sections
that alternated between levels for maximum stability.

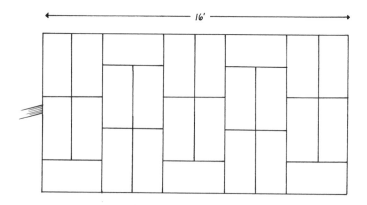

Another, simpler method stacked outside bales in one direction and an inside row in the opposite direction. Though this method was faster and easier, it was also much less stable.

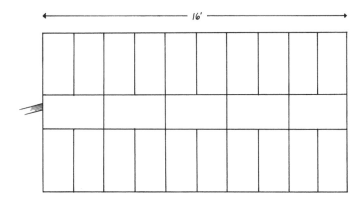

ter failed to make good knots several times in a row, Art would have
to climb the rear of the baler and readjust the knotter settings, hoping for better results.

The process of making a bale of hay was fairly simple (for the
baler compressed loose hay into the hard shape of a bale): Alfalfa was
cut with a mower pulled behind a tractor. The swath of mown hay
on the ground was raked with a side rake into a high, fluffy, quick-
drying windrow with the leaves contained mostly on the inside. The
baling machine had a continuous-feed pick-up that lifted the
windrow onto a conveyer, which took the hay into the baling chamber. There a knife on a plunger sheared off a slab of compressed
hay—akin to a sizable slice of bread. Between twelve and fifteen such
slices collected in the bale chamber, where they were bound into a
bale with twine. The baler kept most of the alfalfa leaves, rich in protein, intact within the bale of hay.[28]

After the bales were transported on the hayracks, they had to be
unloaded into the barn. This step—getting the bales to the loft of the
round barn—proved to be more difficult than the Rongens expected
when they bought the baler. Art, a conservative and frugal man, had
not bought a bale elevator at the same time he acquired the baler and
the side-delivery rake. As a result, he and Doug had to use the old
hay-carrier system, which actually involved *more* work than getting
loose hay to the loft. First, Art had to lay the hay slings on the
ground by the barn, and then he and Doug stacked a dozen bales on
each one—five directly on the sling ropes and three or four layers of
bales onto that foundation. Then they hoisted the loaded sling up to
the peak by the hay door, using the tractor for power, guiding the
bales along the carrier track. They dumped the bales by pulling on
the release rope and then repeated the process. It took nine or ten
sling-loads to unload a hayrack with an average load of 100 bales.
The obvious problem was that Doug and Art had to move the hay
one more time than the old nonautomated system required. Ironi-
cally, stacking the bales on the sling made the process take longer
than it took to put loose hay into the hayloft. So much for efficiency.

Once the bales were in the hayloft, both Rongens went up and threw bales to the extreme edges of the loft and stacked them up, seven high, against the walls. Then they let the remaining bales drop from the sling until the hayloft was filled to capacity; they usually stacked those final bales in the cool of the evening or on a rainy day, when the hayloft wasn't sweltering.

After they had completely filled the loft, Art and Doug stacked extra bales outside the barn but near to the cowyard for ease of feeding. Because these bales, stacked in the shape of a large rectangle or square, needed protection from rain and snow, they had to cover them with large sheets of black plastic. Unfortunately, this material drew heat from the sun's rays, causing moisture to develop underneath it; the resulting condensation caused the top bales to spoil. The bottom layer of bales in an outside stack also rotted because the bales absorbed moisture from the ground. Farmers either accepted the loss of a layer or two of bales or invested in bale sheds, costly but effective shelters with steel roofs. Eventually, Art abandoned the black-plastic covers and topped off the outside stack with loose hay, mounded so the bale resembled an old-fashioned domelike haystack, which readily shed rain.

In that summer of 1959, the Rongens baled all the alfalfa and clover hay they had grown in the fields at the homeplace. They also baled some of the wild hay on the prairie, this time using the new tractor, instead of horses, for power. When winter came, Doug fed the hay to the cows, removing three of the twelve to fifteen "slices" in each bale and tossing individual "servings" into each cow's stall. He had to perform one more procedure with haybales—cutting the twine with his pocketknife, hanging the strands on a nail in the wall, and collecting them after each feeding. If the cows chewed on and swallowed the strings, their stomachs would be upset. And since each cow had four stomachs, that could mean major trouble.

Another potential problem with the new baler was that it not only picked up the hay in the windrow but also everything else that had been raked together into the row, including bits of wire, nails, and other pieces of junk, which farmers lifting the hay with pitchforks would notice and discard. When feeding the bales to the cows, it was

fairly easy to spot the bigger pieces of refuse and throw them away.
But frequently the cows ingested small bits of metal. Cows with
stomachs damaged by so-called Hardware Disease obviously produced less milk. Eventually, veterinarians developed magnets shaped
like slim capsules that the cow could swallow and that would attract
the metal objects and hold them in the stomach as long as the animal
lived.

All in all, the summer of 1959 represented an irreversible turning
point for the Rongen family and their farm. And the new baler and
tractor were tangible evidence of an extraordinary change both for
the Rongens and for American agriculture—the replacement of muscle power in farming by engine power. The Rongens' tractor had performed as it was supposed to, but the baler, with its troublesome
knotter mechanism, had caused some difficulties. Art accepted the
fact that new machines may require adjustments. Yet he questioned
whether the new system of making hay into bales was any great improvement over the old techniques, which he, his father, and grandfather had practiced for years.[29] By now, though, Art had sold his
team and owed on his new machinery, so there was no turning back
to the horse-powered method of making hay.

In the years after World War II, three major changes combined to
make the old diversified farming tradition seem obsolete: the increased mechanization of the farm; the introduction of modern chemical fertilizers, herbicides, and pesticides; and improved plant and animal breeding. As farmers became more efficient and productive and
the United States needed fewer of them to produce ever-greater
quantities of agricultural products. The machine method and the
chemical method were quickly supplanting the old "armstrong"
method of raising crops and livestock. Not surprisingly, the 1960s
saw the end of the dairy-farming tradition for the Rongens of Fertile,
Minnesota.[30]

After two seasons of haying with the new baler, Art knew that
Doug, soon to graduate from Fertile High School, would no doubt
leave the farm. He also had to admit that making hay by himself with

the modern bale system would involve more work than putting up the loose hay with his tractors. So, having learned that he had bought a baler with the worst possible knotting mechanism in those days, he parked the frustrating piece of equipment in the grove. Disenchanted with International Harvester—and flying in the face of a Rongen tradition—Art then bought a Farmhand hay loader from a dealer other than Solon Gullickson.

The new loader attached to the front of the tractor and had a dozen nine-foot-long tines for picking up hay; a hydraulic lift raised and lowered the tines. Art gathered up loose hay with the loader and then dumped each full load into a "stack frame," which was made of steel bars and measured twelve feet high by twelve feet wide by twenty feet long. After Art had filled the stack frame to overfull (to allow for settling), he removed the frame by unlocking a gate in the back and pulled the frame with his tractor to the next location, ready for more hay. These resulting stacks resembled huge loaves of bread, instead of being mound-shaped like the traditional haystacks.[31]

This system meant that Art had to hire a neighboring farmer to bring over a mechanized stack mover to haul the huge stacks to the barn. Once the stacks were outside the barn, he had to break them up to get the hay into the barn for feeding the cows. Even though this method signaled the end of another tradition—storing hay in the loft—Art Rongen was forced to adopt it because Doug was no longer available to help him. Right after graduating from high school in 1962, Doug began working as a carpenter in Grand Forks, North Dakota. In 1966, when he was a prime candidate for the Vietnam War draft, he enlisted for a two-year hitch in the Marines. He received orders for combat duty in Vietnam. When his plane landed in Okinawa, he was among twenty-five Marines fortunate enough to be reassigned to duty there.

Upon returning to civilian life in 1968, Doug came home to Fertile and resumed farming with his father. Together they faced several unavoidable questions—questions farmers all over the state were then confronting: What type of specialized farming was the right choice for them in their particular area? Should they expand the dairy-cow operation by building a big new barn with a gutter cleaner and a new silo? Should they abandon dairying and buy into a feedlot system for raising beef cattle? Should they try to become cash-crop farmers, growing corn, soybeans, or sugar beets? The reason behind the Rongens' ultimate decision was tersely summarized by a Redwood County farmer, Clarence Weber: "You either had to get into it big or you had to get out" of dairying. Weber, and the Rongens, got out of dairying.[32]

They lacked the manpower to make enough hay for a larger dairy operation and the financial wherewithal to replace the round barn, unsuitable for modern dairy farming because of its odd dimensions and dirt floor. What was more, their land was not good enough for sugar beets, and they were situated too far north to make much money with soybeans or corn. The best option appeared to be to give up the dairying operation—which had supported the Rongen family for four generations—and begin raising beef cattle.

Yet another Rongen tradition succumbed when, in 1969, Doug married Bonnie Buck, a young woman whose heritage was Irish and Swedish and whose religion was Roman Catholic. For the sake of family unity, Doug gave up his membership in the Little Norway Lutheran Church and, with Bonnie, joined St. Joseph's Catholic Church in Fertile.[33]

Although Doug's marriage shattered an old Rongen family custom, the young man continued to farm the land with his father. In 1972 Doug bought the farm from Art; he also bought and rented more land from neighbors who were giving up farming. That same year, he and Bonnie moved into his parents' farmhouse, and Art and Hilma moved into a mobile home just fifty yards away. When Doug and Bonnie began having children—Wade in 1973, Aaron in 1975—they were grateful that grandparents were close by. Hilma died in 1977, the same year their daughter, Amy Jo, was born. Grandpa Art, now a grieving widower, was nonetheless cheered by her birth, which he called "God's gift" to the family.

Doug Rongen made hay for his beef cattle from alfalfa fields near the homeplace and turned the family's prairie property over to pasture for grazing. He began making the kind of large, round bales that have become common throughout rural Minnesota; these are often left in the fields and road ditches during the summer and fall and later hauled to a farmstead for feeding to livestock. These distinctive bales have become the late-twentieth-century equivalent of the large haystacks of earlier generations. Their shape suggests a massive portion of shredded-wheat cereal. And their presence on a farm means that the hay they contain will be fed to beef cattle in outside feed-bunks rather than to dairy cows inside a barn.

A round baler winds more than a thousand pounds of hay into a distinctive cylindrical shape. Compared with the standard square bales, these are behemoths. The baler then wraps a single twine string around each gigantic bale and drops it out the rear of the machine. The outside of the bale turns brown or gray through the weathering action of rain, sun, and wind, but the inside remains green and nutritious. The baler winds the hay so tightly that moisture cannot easily penetrate the bale. The bales must be moved to the edge of an alfalfa hay field after the first hay harvest in June; otherwise, their great weight would crush the plants underneath, preventing a second harvest.

Doug Rongen's first round baler had imperfections. For instance, if he put the windrow directly in the middle of the pick-up mechanism while baling a field, the machine would produce uneven bales—too large in the middle and too small at the ends, or vice versa. He learned to vary the input, driving from side to side—a little to the left of center, then a little to the right of center—in order to make evenly proportioned bales.

The manufacturer of Doug's baler gave no instructions regarding how to stack the large bales, so he and other farmers had to experiment. If he stacked them end to end in a long row, with only the ends touching, very little of the hay rotted. But if the sides of the bales touched, rain would collect in the cracks between them, causing spoilage.

This method of baling hay may be touted a one-man operation, but moving the enormous round bales to the farmstead without assistance is punishing work. Doug used a tractor equipped with a hydraulic loader to lift the bales onto a hay wagon, which he then pulled home behind the tractor. Other farmers employed small Bobcat forklifts—like those used in warehouses to lift wooden pallets—to haul the bales from the field to the farm. Once there, the round bales are satisfactory for feeding beef cattle in outdoor cribs, where cattle can bite off chunks of hay, but dairy farmers tend to prefer small, square bales that can be conveniently stored in the hayloft and fed to the livestock inside the barn.

Doug Rongen (who today owns 540 acres of land, including 150 acres of alfalfa for hay) built a pole barn as a cold-weather shelter for his beef cattle; in summers, the animals are in pastures. A pole barn costs much less to build than a dairy barn. It consists of a simple frame made of long poles, with metal siding and a steel roof. One side of the building (typically the south side, so winter sun can shine in) lacks a wall; that way, manure can be cleaned out by means of a tractor and loader. In many ways, a pole barn is a large version of the basic sheds that Minnesota's pioneers built to shelter their livestock from harsh winter weather. Modern pole barns are cheap, simple to build, and easy to maintain. What they lack, of course, is a hayloft.

True to the Rongen family tradition, Doug maintains a multi-generational farmstead. His father Art, now in his early nineties, still lives in his mobile home on the homeplace and tries to help with whatever farmwork he can manage. Doug, like many other present-day farmers, has had to find outside work to earn enough to pay his mortgage and property taxes: he grades the gravel roads for two townships and also does maintenance work at a local nursing home. His wife Bonnie is the postmistress of the nearby town of Beltrami.

Will the farming traditions of this Norwegian-American family endure as they have for the German-American Marthalers of Meire Grove? The answer for the Rongens lies in the youngest generation, which must decide which, if any, of its members will take over the farm operation. Doug's son Wade now works and ranches in Montana and may remain in the West. Wade's brother Aaron lives at home but works for a local tree nursery, owned by another family of Norwegian extraction. Daughter Amy might carry on the farming tradition—if she marries a man who is also willing to work the land. But if none of the Rongen children decides to buy the farm, the land will probably be rented out to a neighboring farmer and, eventually, sold.

Even though the changes around Fertile since the days of Johannes Rongen are incalculable, two of the area's most distinctive landmarks remain. The Little Norway Church has half the members and twice the gravestones as it had fifty years ago, but the peak of its spire remains taller than any other building for miles around; Art still attends services there. And the Rongens' singular round barn is still in good condition, though the inside is empty except for memories.

The Rongen round barn as it appeared in July 1996. Note the new hay rake in the foreground.

Remove not the old landmark; and enter not
into the fields of the fatherless.

PROVERBS 23.10

After my dad died, I drifted for a number of years. During my senior year of high school, my choir director, Jon Wittgraf, encouraged me to pursue music. I had a good voice and he helped me get into the University of Minnesota Summer Musicians Project, where I learned a few tricks of music theory from the peerless Vern Sutton. But when I declared a music major at Gustavus Adolphus

Dad aboard a Navy ship in the Pacific
during World War II, about 1944.

the following year, I found out just how little theory I knew. I knew even less about being in college. I dropped out of Gustavus Adolphus and its famous Concert Choir midyear. A few years later, with a better understanding of where my interests lay, I went back to school at Bemidji State University to study history.

When I would come home over vacation, my brother Larry would always have me draw the spots of the Holstein calves on their registration papers, just as Dad always had, and every night I would go to the barn for evening milking to help Larry carry the milk to the bulk tank. We would talk as he worked. I told him about my classes at college and asked his advice, now that I was dating my future wife, Dianne.

But still I drifted, fatherless. After I got my B.A., I went to Luther-Northwestern Seminary in St. Paul for a while, thinking that I might become a pastor; but I couldn't stand the bustle and traffic of the Twin Cities, and I missed Dianne, who was still attending Bemidji State. So I went back to Bemidji and got my social studies teaching degree. For a short time I taught high school at Belgrade, just eight miles from Meire Grove, then spent eight years at Staples High School. Soon I was restless again, after losing my brother, and wanted to get out of the Midwest, so I went back to school, completing my M.A. in history at the University of Vermont. Ever since I read those Landmark books about Ethan Allen and his exploits in the American Revolution, I had dreamed of exploring in those Green Mountains. I wrote about the poorhouses of the area, remembering how my dad pointed out to me the Redwood County poorhouse when I was young. But when it came time to choose a thesis topic, I wrote on the history of a Minnesota logging town, Buena Vista, and its frontier hotelier, John Wesley Speelman (my children's great-great-great-grandfather); I was feeling the tug of my Midwestern roots. I got my doctorate in U.S. history from the University of North Dakota, concentrating on regional and social history of the twentieth century. I wrote about "Prairie Paupers," the history of poor people and welfare in North Dakota, again my father's suggestion.

Gradually, I have come to see that my interest in history stems from the need to assemble a story from scattered fragments in my own life. I remember how Dad told us about sailing west in the Navy during the war, how his ship was almost hit by a kamikaze plane in the Battle of Okinawa. He kept coins and currency from Occupied Japan, China, and the Philippines in a cigar box in the upstairs dresser. I remember how I liked to hold the coins, to look at the strange characters on the paper money, how I liked to see his photographs of faraway places like Kodiak Island and the cities of China. I never got to ask Dad all the questions I have now about his time in the war, and I see now that the things we never got to do together affect me as deeply as the things we did do.

When he would go pheasant hunting, I was still to young to go along, though now I take annual hunting trips into the woods to get ruffed grouse. I never played catch with my dad, but I held his funny-looking 1930s glove and batted with his Babe Ruth model Sears, Roebuck bat. And though I hated hoeing, weeding, and picking in my dad's small orchard of apple trees and rosebushes, last year I planted two apple trees and a rosebush of my own. But I never forget: my dad was a farmer and I will never be one. Though I grew up on a farm, I never had any desire to become a farmer, no inclination to do farmwork. I always preferred having books in my hands rather than tools.

Yet, as I got halfway through working on this book, I had to come to grips with the old feelings that I had somehow betrayed my dad by leaving the farm. Those feelings only deepened when my brother died and I did not return to take up the business. When my brother's farm equipment was sold at auction, I could not bring myself to go. I should have gone, for it was like a funeral for the farm—for the haylage machines, for the tractors, for the New Holland baler and rake— and I felt ashamed.

When I examined the five families for this book, I came to envy the Marthalers and the Rongens because both have kept the farm in the family. I began to think that I might have taken Larry's place as the farmer after he died. Even though I was totally unsuited to the task, I vaguely felt that I had failed in my duty, being the oldest son. Even worse, no one really expected me to come back to the farm. Yet, in the end, I came to understand that I could not take over the family farm. The Marthaler family and the Rongen family have had their fathers there to help them. For me, it was too hard to become a farmer without my father to teach me.

A few years ago, I was able to move back to Minnesota, where I belong— closer to my mom, to be near her in the place of my dad and in the place of my brother. I stand beside her at relatives' funerals, finally taking the mantle of the oldest son, the man of the family, albeit the one who did not take over the farming operation when my brother died. Yet, I see now that my boyhood on our dairy farm enabled me to write this history of haying. My memories of Dad and Larry and the work they did guided my research. I have tried to honor my heritage by helping to preserve a bit of the everyday life on the farm. The memory of my father helps me in my work as a teacher and writer; I still tell my history classes about his close call with the kamikaze plane at Okinawa. And the memory of my brother reminds me to be gentle and patient. I hold on to him by writing his story. I have had a number of dreams about each of them, all good ones, and I have held dear the line from Song of Songs (8.6): "Love is as strong as death."

Not one of my brothers and sisters is a farmer now, so the 110 acres that remain of the farm at Morgan are our last connection with our heritage. I claim to be guiltless over leaving the farm. I blame my lack of vision on my dad's death and my brother's, but those who give up their farms without a fight, without seeing if they could have kept it going, always live wondering what might have been. Though I will never be a farmer, I will do my best to help my family hold on to that little plot. If we can preserve at least that property, that land my grandfather bought and named as his, our, homeplace nearly a hundred years ago, maybe we can hold together well enough as a family.

Blue Silos on the Prairie
HAYING WITH SWATHER, TRACTOR, AND CHOPPER
Larry Hoffbeck, Danish American
Morgan, Redwood County, 1984

My brother Larry always wanted to be a carpenter, to work with wood—cutting it, crafting it, shaping it into cabinets, bookshelves, even buildings. Woodworking was his joy, and he wanted it to be his livelihood. Although he was the oldest, he hadn't been groomed to assume control of the farm on that distant day when our dad would retire. There were five more boys—me, then Jeffrey, Dana, Chris, and John. Dana showed the most interest in the place and would be old enough to take over the farmstead by the time our father stopped farming, so Larry was free to choose his field of work on graduation, and what he wanted to do was carpentry.

Born in 1950, Larry was a so-called baby boomer, but he learned how to work on the farm as if he were from Dad's generation. He knew how to feed the cows, pitch manure from the calfpens, and hoe the cockleburs and other weeds out of the soybean fields. And he knew how to make the most of his free time. On those long summer nights, we played softball in the large yard south of the house, swung on

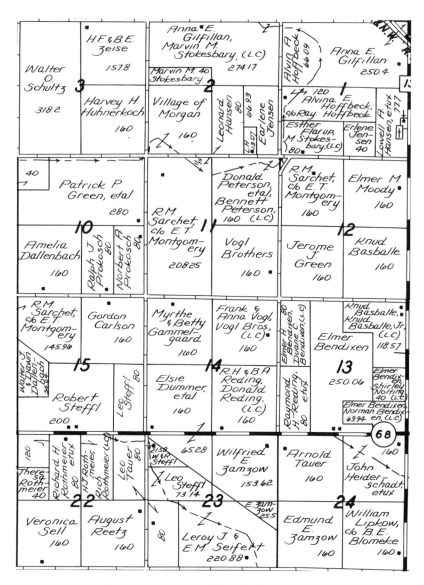

Our family farm, shown here under "Alvina E. Hoffbeck c/o Ray Hoffbeck,"
sits near the northeastern corner of Three Lakes Township, Redwood County,
as shown in this detail of the plat, 1977.

Page 139: Hay swather like those used in the 1980s.

ropes in the hayloft, shot baskets through a homemade hoop. We raced our bikes on the gravel driveway and reenacted Civil War battles, shooting at each other with cap pistols, feigning death by falling down and counting to ten.

Because Larry was skilled with hammer and nails, he also loved building tree houses. One of his most ingenious, constructed in the east part of the grove in a tree over the cowyard, had a four-pane window and a floor with a trapdoor. In the west portion of the same grove, Larry built a dugout house, underground. None of us ever worried that the roof of reused two-by-fours might collapse, though we knew we'd shoveled enough dirt on top to suffocate us under an avalanche of dust and lumber, if it did. It was simply unthinkable that one of Larry's creations would be anything but safe and solid.

Soon after graduating from Morgan High in 1968, Larry became a carpenter's assistant to Christ Kruse from nearby Wabasso. A devastating June tornado had swept through the neighboring town of Tracy, destroying homes along a four-block swath and killing seven people. Larry worked with Kruse to build secure shelters for the storm's victims and to repair buildings. He liked the work of helping rebuild Tracy in part, I suppose, because it seemed light-years away from other scenes of destruction in 1968. The North Vietnamese had attacked U.S. troops in South Vietnam during the Tet Offensive; Robert Kennedy and Martin Luther King Jr. had been assassinated; rioting and protests erupted in cities and on college campuses across

(L to R) ABS technician talks with John, Jessica, and Larry, July 1982.

the country. Knowing he was eligible for the selective service, Larry feared that the war might claim him, too.

Other young men from Morgan had already gone to Vietnam, and two of them, Paul Basballe and Tom Kiergaard, were from the same country church that we attended. Another neighbor, a German Catholic named Ron Kerkhoff, wrote to the *Morgan Messenger* from Vietnam, telling everyone that he had "lost a lot of buddies during the fighting." He admitted that "everybody" in his outfit was "half-scared to death including me." No wonder Larry suspected that he too would soon be called up. What he didn't anticipate, however, was a different kind of tragedy.[1]

In late November of that year, our dad was still trying to complete the work of gathering in the soybean crop in the field just north of the family farmstead.[2] His early-1950s McCormick combine-harvester still worked well enough, but its safety features were out-moded. The combine got its power from the tractor via a PTO shaft, and old-fashioned steel shields were all that formed a flimsy hood over the rapidly whirling metal knuckles. Dad leaned over and his bulky coveralls got caught in the revolving PTO shaft. Its whirling knuckles grabbed him and fatally embraced him. He was pronounced dead on arrival at the Redwood Falls Hospital.[3] He was fifty years old, and he left Mom with nine children, all but one still in school. Larry, as the oldest son, volunteered to run the farm. Now, he wouldn't have to go to Vietnam. Nor would he pursue his lifelong ambition of becoming a carpenter.

The farm had been kind to our family until that day in 1968. My grandfather, George Hoffbeck, had emigrated from Denmark in 1912, at age twenty-six, and eventually bought a farmstead just five miles northwest of Morgan. It was near a large rural estate called Gilfillan, named after the land baron Charles D. Gilfillan, who owned an empire of ten thousand acres of prime cropland in Redwood County. While no empire, the black loam on our 166 acres of flat prairie-land was among the best on earth—three to six feet deep, atop clay subsoil—so it was easily plowed but held its moisture well.[4]

The Hoffbeck land adjoined that of other Danish immigrants, families whose names all ended in "-sen"—Bendixsen, Hansen, Knudsen, Rasmussen, Jensen, and Simonsen. Our family name sounded German because the family came from Schleswig-Holstein, an area that had once been German-controlled, but the Hoffbecks considered themselves true Danes. Together with other Danish immigrants, they built a Lutheran church on the road just a mile across the field from George Hoffbeck's place, near the country school. They called it Bethany Lutheran Church, and George sang in the choir. Except for one or two Norwegian families nearby, the area around Morgan was inhabited almost exclusively by Danes and Germans. German Catholics had built St. Michael's Church in town, and the German Lutherans built St. John's Church just two blocks away.

In 1917, George married Ella Nielsen, a young Danish woman who played the organ at Bethany Church. The following year, the first of their eight children was born, a boy they named Raymond, my father.

From his birth until his death, Dad never left the farm, except to serve in the Navy during World War II. He was a storekeeper on a ship in the Pacific that was narrowly missed by a Japanese kamikaze during the battle for Okinawa. When returned home, in the mid-

My uncle Norman using a sling to put up hay in my grandfather George's barn, 1942

1940s, he started buying the 166-acre farm bit by bit from his parents, who by then had moved into a house in nearby Redwood Falls. Dad also bought his father's farm equipment, most of which was for horse-powered farming but which included a gasoline-powered tractor purchased in 1940. Shortly after Dad began farming he knew he wanted to get married and start a family. He was thirty years old when he met Alvina Engel, one of the German Lutherans, at a social at St. John's Church. She was eighteen in 1948 when she became his bride, and the couple had the first of their nine children, my sister Janice, a year later. Grandpa George had taught his eldest children, including Dad, to speak Danish, but he was intent on becoming as American as possible, having learned English at the Danebod School in Tyler, Minnesota. None of us, out of forty-two grandchildren, ever learned our grandparents' native tongue, but we did learn their work ethic.

We boys were expected to do the chores in the barn and around the farmstead, while Dad did the actual work of planting, cultivating, plowing, and harvesting the crops. Together we all participated in bringing in the hay, which provided enough winter feed for twenty cows, plus numerous immature heifers and calves. We made about five thousand bales of hay each summer during the 1960s, getting high-quality hay from the hayfields and some grass from the road ditches. The boys worked in the hot hayloft, stacking the bales to the rooftop. The oldest girls, Janice and Ane Marie, sometimes helped

What little European culture my parents did retain mostly had to do with food. For example, Mom learned from her mother-in-law how to make a sausage called *Medista Puls*—pork and spices in intestine casings. She also baked hard little cookies called *Pfeffernusse*, which had to be soaked in coffee or milk in order to be eaten. Dad showed Mom how to use leftover roast beef, cubing it and heating it in milk and pepper in the Danish style. Another Old Country custom they adopted was to make a humble supper of hot milk and chunks of bread. But only Raymond used bread crusts to wipe up gravy and mashed potatoes or other foods from his plate, the way Danish peasant farmers had cleaned their plates for generations.

drive the tractor that pulled the baler. But Dad did almost all the work on the hayrack in the hayfield, making his loads five layers high.

In 1967 he completed an addition to the barn so he could increase the dairy herd from twenty-four to forty cows. And he modernized the structure by installing an automatic gutter-cleaner that conveyed the manure out the back door into the manure spreader. This meant we no longer had to use a manure fork and a shovel to pitch it into a litter carrier, which ran along a ceiling-mounted track the whole length of the barn.

After Dad died, Larry accepted responsibility for farming the family's land. Mom continued to take care of the rest of us despite her grief, but without Dad, we were dependent on Larry to run the farm and thus to provide a livelihood for the family. Although he was just eighteen years old, he was suddenly responsible for the seven children still at home—John, age one; Chris, four; Brenda, five; Dana, eleven; Jeffrey, fourteen; me, fifteen; and Ane Marie, seventeen. One of the first things Larry did was to get rid of the combine that killed Dad. Our neighbors, all fellow Danish-American farmers, used their more modern combines to help us finish the harvest Dad had been rushing to complete. Even afterward, Larry always hired a neighbor to combine the beans and pick the corn rather than use a combine himself.

Standing just five foot ten and weighing only 160 pounds, Larry was not a big man, but he was sturdy and strong. According to family lore, he once ate fourteen ears of sweet corn at a single sitting. Like Larry, our farm was of modest proportions, consisting of 166 acres. During the earlier period of diversified farming, 160 acres was considered to be sufficient land for a whole family, but by the late 1960s, the industry was changing. A small acreage was not enough to provide a large income with cash crops such as corn and soybeans, but it could provide enough feed for a moderately sized dairy herd of forty to fifty cows. Larry had to decide whether he could afford fully to modernize his relatively small farm. Was 166 acres enough land to sustain an upgraded dairying operation?

[handwritten margin note: is this still today?]

Larry enjoyed two advantages as a beginning farmer: his father had maintained a herd of forty high-quality black-and-white Holstein cows; and his barn and animals were certified by the State of Minnesota to produce Grade A milk, which meant that the operation commanded the highest bulk-milk prices. Inspectors visited Grade A farms regularly to check that farmers had installed and maintained stainless-steel bulk-milk tanks and stainless-steel milking equipment—all disinfected after every milking—and to be sure the barns and cows were clean. Grade B farms used old-fashioned milk cans and were not as meticulously cleaned.

The 1960s brought specialization to Minnesota's farmers, many of whom decided they did not want to invest heavily in dairying equipment and expand their lands and barns. Some dairy farmers left their farms altogether or turned to raising beef cattle or cash crops because milking cows twice a day was simply too arduous and time-consuming. Because many believed that the "outlook for fair returns for labor and investment [was] not good," dairy farmers left their farms during the 1960s at twice the rate of the past decades.[5] In 1968, our dairy farm was one of 56,039 in Minnesota; this total represented a twenty-nine percent decrease from 1962, when there were 78,482. While the number of dairy farms declined, the number of cows stayed roughly steady, at around 1.2 million. At that time, Minnesota was first in the nation in the amount of butter produced and third in total production of milk.[6]

Larry decided to specialize in dairy farming since Dad had already begun upgrading the operation, but he needed help in getting started. Although he had learned some aspects of farming, he had to learn more about tilling the soil and caring for the livestock, and one of his teachers was Uncle LaVerne Dahmes, who lived just two and a half miles north of our place. Uncle LaVerne also had a forty Holstein herd and a 160-acre farm. His son, Michael, had always been Larry's best friend, so LaVerne, Larry, and Michael already enjoyed a close relationship.

Uncle LaVerne understood that all of us, Larry most of all, needed him to be a good uncle and help with running the farm. LaVerne had

had a close relationship with Dad; he mourned his passing and lived with the knowledge that his own family could have been going through this suffering, for LaVerne had narrowly avoided death in a farm accident himself. In the early 1960s, he had climbed up into his silo soon after filling it with chopped corn silage, and he had been nearly suffocated from silo gas in his lungs. He managed to get to clear air in time to save his life but his lungs suffered permanent damage. Uncle LaVerne took care of us the way Dad would have cared for LaVerne's family.

When he began farming, Larry "had no idea of how to plant corn" or when to plant it. Uncle LaVerne taught him "the calendar" for planting corn and the techniques of modern farming: when to apply fertilizer to augment the natural cow-manure fertilizer, how to spread herbicides to kill the weeds in the soybeans, and how to spray crops with pesticides to kill corn borers or grasshoppers. He also taught him the fine points of haymaking—when to cut and bale it and how to store it.[7]

To help him run the farm and especially to help him make hay, Larry depended on me, Jeffrey, Dana, and even Chris. During the early 1970s, we did our haying with equipment inherited from Dad, but Larry hated the inefficiency of making square bales of hay, having to repeatedly handle the bales.

One day, Larry grumped about the tedious process; Chris, who couldn't have been more than eight then, said that he didn't mind driving the tractor for baling hay.

"I like the smell of the hay," Chris said.

Larry said, "It smells like *work*."

Then we brothers began, one by one, to graduate from high school—first me in 1971 then Jeffrey in 1972. I went to college at Gustavus Adolphus, then Bemidji, and Jeffrey to Willmar Vo-Tech. When Dana graduated in 1975 and went to Jackson Vo-Tech to learn the electrician's trade, the baling became especially taxing for the few family members left at home.

Larry hired an operator to make large round haybales for two haying seasons, but he found that the five-foot-wide bales were diffi-

cult to move from the field to the shed, and were too large to fit
through the hayloft door. Furthermore, it was so difficult to break up
the big bales in order to feed them to the cows inside the barn that
Larry eventually decided to let the cows out to eat from feeders in the
cowyard. The large bales worked well for farmers who raised beef
cattle, which could live in pole barns with easy access to outdoor
feeding rings, but they did not work for farmers with conventional
dairy barns. Before long, Larry went back to baling all his hay with
the New Holland baler.[8]

By 1976, he decided to try a more modern haymaking system.
Some Minnesota dairy farmers were now using a type of feed called
haylage, which was hay that had been chopped and stored in a silo. It
differed from conventional hay in that more moisture remained in its
stems and leaves; only fifty percent of the moisture had been allowed
to dry in the field, so when haylage was stored in a silo, it fermented,
acquiring a different taste and smell than regular hay. It was similar
to corn silage—chopped corn and cornstalks stored in a silo—but it
had the food value of alfalfa: high protein with enough roughage for
a cow's digestion. Haylage had forty to fifty-five percent moisture,
while corn silage had fifty-five to seventy percent.[9]

Haylage could be stored in the kind of concrete silo that had been
popular in Minnesota since 1910. These silos stored corn and hay
silage equally well, but they did allow oxygen to come in contact with
the silage, resulting in a slight degree of spoilage. The up-to-date al-
ternative to a traditional concrete silo was called a "sealed storage"
silo, made of glass-fused-to-steel sheets held together with steel
bolts. The interior glass coating kept most oxygen out of the silo to
preserve the haylage better than the old silos. This new type of silo,
which resembled a giant Thermos bottle in form and function, was
available in various colors—the dark blue of the Harvestore silos, the
dark green of the Cropstores, the light blue of the Sealstore silos. All
these sealed-storage silos had huge capacities for haylage. On the
farm landscape, a haylage silo was like a giant blue or green hayloft
tipped up on its end. They were filled from the top and unloaded from
the bottom.[10]

In 1974 Larry attended an informational meeting about the advantages of haylage stored in blue silos, and he was "all hepped up" about getting one.[11] The haylage system promised greater efficiency: if alfalfa was quickly cut, chopped, and blown into the silo, the precious leaves, rich in protein and nutritive value, would be less likely to fall off or blow away or become spoiled during the haymaking process. After deliberating and examining the advantages of haylage stored in the big blue silos over haylage stored in concrete silos, he decided to purchase a large blue Harvestore silo. It was a major expenditure—$35,000 for a silo twenty feet in diameter and seventy feet tall. A concrete silo of those exact dimensions would have cost considerably less, about $20,000. (Stunned by the high price and skeptical about their effectiveness, some dairy farmers considered the sealed-storage silos to be a luxury and not that much better than the concrete type.[12])

To help bankroll his move toward a more modern way of making hay, Larry sold his bale-stacker machine to Uncle Alvin, from Tracy, for $3,600 and used the money to buy a used windrower machine. Pulled by a tractor, this device cut the hay, squeezed out some of its moisture between rollers, and then placed it in perfect windrows in the field—all in one operation. It had a twelve-foot-wide cutter head which cut twice the amount of hay that Larry's old mower could cut with its six-foot wide cutter arm. Although this machine was technically known as a windrower, Larry and others called it a "swather," for the swath it cut through the fields.

Progress meant other major expenses as well. Larry found it necessary to purchase a Badger field chopper that would cut up the haylage for the new storage system; he borrowed from his local banker to buy a used model for $7,500. In order to transport the haylage from field to Harvestore, he bought two new silage wagons called "forage boxes," each of which cost $4,000. The classy wagons had built-in unloading systems for speedy movement of the haylage from field to silo. Each one was twenty times more expensive than a plain old hayrack, but could be handled by one man.[13] In order to expand

and modernize at this rate, Larry had to become an agribusinessman; that is, he had to raise capital and maintain a debt load.

To pay for all the machinery, Larry needed to increase his income from the cows, because the best way to make sure he had a substantial cash flow was to get more milk. The federal government had increased dairy price supports in 1973, and milk prices were good through the rest of the decade. Haylage was advertised as a means to get better feed for the cows and, as a result, more milk. The larger his milk check, the faster he could pay off the loans he had assumed in order to buy the whole system.[14]

To increase milk production, he built an addition to the north side of the barn in 1977, allowing room for ten more cows. He had to purchase the materials, but he did much of the construction work himself, making wooden forms for the concrete work and building the roof and shingling it.[15]

Larry also bought a pipeline system to get the milk directly from the milking machines to the refrigerated bulk-storage tank. The pipeline did the work his brothers had done by hand in the 1960s. It cost money, but it saved labor.

*Larry atop his
Harvestore silo,
late summer 1977.*

In 1973 Larry married Wendy Sandmann, a Lutheran of German background who also had graduated from Morgan High School. At first the young couple lived in a farmhouse they had bought just half a mile from the farm, but before long their family outgrew that space. Their first daughter, Jessica, was born in 1975 and their second, Melissa, in 1978. To accommodate his growing family, Larry bought the old farmstead from Mom. He moved her into the smaller farmhouse and moved his family into the house where we all grew up.

When Larry and Wendy decided they wanted a new farmhouse, they sold the old house to someone who moved it into Morgan and set it on a new foundation. Then they borrowed about $50,000 in order to build a split-level home. Larry, who had so wanted to be a carpenter, welcomed the opportunity to build a new home from the ground up rather than live in one that had been remodeled and added to several times. Larry did much of the woodwork in order to hold down costs, including the kitchen cabinets—all of oak, his favorite wood. He also built and outfitted a workshop for himself in the basement, and there he built shelves, cabinets, and other items in his scant free time. And when they needed a dining-room/laundry-room addition, to pay for it Wendy worked for a while at a Control Data plant in nearby Redwood Falls.

Larry arranged financing with the Farmers Home Administration (FHA) for $82,000 to buy sixty-six acres of the farm from his mother in 1978. She gave him a good price of $800 per acre, considerably less than the going rate of $2,000 per acre for choice southwestern Minnesota farmland. The FHA was the lending agency for many of his neighbors, who were also modernizing and enlarging their operations as agribusiness propositions; the agency made loans to farmers who could not get any more money from other sources.[16]

By becoming parents (of three children now, including Landon, born in 1981), Larry and Wendy added to their mounting financial responsibilities. Debt in itself is not necessarily a bad thing, but debt in the late 1970s and early 1980s could be debilitating. The Arab Oil Embargo of 1973 raised oil prices and caused an inflationary spiral in the American economy. Farmers were hurt more by the higher costs of fuel for tractors, petroleum-based fertilizers, and farm chemicals, than they were helped by nominally increased farm-commodity prices. Essentials for baling, such as baler twine, more than doubled in price; for example, a ball of twine that cost $9 in 1973 cost $23 in 1974. The secondary oil shock of 1979, after the fall of the Shah of Iran and the subsequent cutoff of Iranian oil to the United States, further increased inflation and elevated interest rates. Larry mentally

By the late Seventies, I could sense the old way of life changing on our farm.
Over Christmas vacation, 1977, I took this photograph of the old
mowing machine at Mom's house rusting with disuse.

maintained a running total of his growing debt, and even his mother admitted to worrying about her son owing "quite a bit of money" to the bankers.[17]

Being part of the agribusiness revolution required more technology and Larry responded to the call, putting a phone in the barn's milk room in 1976. He also installed a stereo radio system with extra speakers, so he could hear WCCO-AM broadcasts throughout the building. He tried playing the rock-music broadcast on WDGY-AM or KDWB-AM from the Twin Cities, but the cows were too agitated by the driving rhythms—so he tuned back in permanently to 8-3-0, WCCO.[18]

In 1979 Larry made the fateful decision to buy a second blue silo. The local blue-silo sales representative convinced Larry to acquire another one, telling him that increased storage capacity would enable him to have sufficient feed all year round, rather than having to pay other farmers for additional hay. Larry listened and bought—and fell into a money trap from which there was no escape. The second silo (thirty feet in diameter and seventy feet tall) would cost $70,000, both because it was larger than his first one and because of climbing prices, due to inflation. When the local banker calculated Larry's debt load, he refused to approve a loan for another, much more expensive, blue silo. Skyrocketing interest rates would have made it difficult to repay a loan at eighteen-percent interest anyway.

The salesman then offered Larry another option: a lease agreement on a blue silo and also on an automatic feeding system for an additional $12,000. Also called a brush feeder, because brushes with bristles pushed the haylage along a conveyor track, the automatic system included a computerized scale that weighed the haylage as it passed through and stopped the unloader when it had supplied enough feed. Larry could punch in the number of pounds of haylage he wanted, and the feeding system would deliver the haylage directly to the outdoor feeding bunks or to a feed cart (for wintertime indoor feeding at the mangers in front of the cows). The lease agreement on the silo was for ten years, after which Larry would have the option to buy both it and the feeding system. If he did not buy them, the company would take down the steel silo and remove the automatic feeder.

So Larry signed the agreement, seeing no other way than increased automation to get all his work done. Only after the second blue silo stood on its new concrete slab south of the barn did he discover the limitations of the haylage and the storage silos. Even though he planted more alfalfa on his farmland, he found that he could not get enough haylage in a season to feed his whole dairy herd through the winter and into the new haying season in May; he always had to buy some large round bales.

Ultimately, the experience of owning two blue silos became a nightmare for Larry. Neither unit proved to be as airtight as promised, and he had to have a number of leaks resealed. The repair expenses mounted. Bad weather in the late 1970s meant less hay and other crops than Larry anticipated. Some of his alfalfa fields suffered from winterkill, when excessive cold combined with scant snow-cover to destroy the plants.

By 1980, Larry could only hope that his efficient method of haying would give his dairy herd the superior nutrition necessary for producing enough milk to pay for the whole operation. His fifty-cow herd provided a regular income, and productivity did increase—but not as much as expected. The lease agreement on the blue silo stipulated that the monthly payment be deducted automatically from his milk check. The bank also required that the monthly loan repayments be deducted from the same check. Larry received whatever was left, and in the worst months of the year, that was very little. He was finding that modernization in tough economic times could mean "harder work, less cash, and more debt."[19]

Larry continued to maintain an operating loan account with the local branch of the Norwest Bank corporation. This arrangement allowed him to purchase seed, fertilizer, and farm chemicals, with payment and interest automatically deducted from the monthly milk check from the Associated Milk Producers Incorporated (AMPI), of New Ulm. The percentage payment from the milk check represented dollars Larry never saw, because one corporation serviced the payments of another. This was agribusiness as usual in the 1970s and 1980s.

Larry and Wendy's massive debt burden represented the cost of modernization, of home, farm equipment, hay-storage silos. They had bought their homeplace at a time when land prices for farmland in Minnesota reached all-time highs, and the resulting debts forced Larry to work his frantically to keep the whole enterprise afloat. Then, in 1981, a worldwide recession reduced demand for U.S. farm crops, and prices fell across the board. Larry had borrowed money to mechanize his farm when crop prices were high; now he had to continue paying off his loans when those prices plunged. By adopting new technology, he had invested beyond his means and had sunk into perpetual debt. It was a cruel irony: he ended up working harder than ever in order to pay for a laborsaving method of farming.[20]

Larry was only one of countless U.S. farmers battered by the "worst agricultural crisis since the Great Depression."[21] Some of his neighbors had to abandon farming because they could no longer make their payments. Uncle LaVerne had bought another farm when land cost $2,000 per acre, but he could not produce enough crops from that property to continue the payments, so it reverted to the bank. Many farmers and their spouses in the Morgan area, burdened by debt and inflationary holdovers from the 1970s, took second jobs in town in order to survive the farm crisis of the early eighties.[22]

In 1984, Larry named his property Tri-Oaks Farm, after the three oak trees on the homeplace. Our Grandpa George had planted two of the trees, which represented the first two generations of dairy farmers on that land. The third one had sprouted from an acorn, and Larry had watched it grow and mature as he did. Perhaps it represented Larry himself as the third generation on that land. Oak was his favorite wood for building cabinets, shelves, and furniture, so what better name could he choose for his farm?

In those days, we relied on Larry to be strong as an oak, both physically and mentally. Larry was the one that all of us, brothers and sisters alike, depended on for help and counsel. He was always willing to help, while he still tried to reserve enough time for his own little family—for Wendy and for the kids. But within him was an agonizing frustration with the tiresome reality of the debt that was half-suffocating him. One of the few ways that he could literally rise above the financial crisis was to climb to the top of one of his seventy-foot-tall blue silos. Whenever he put more haylage in the silo, Larry had to seal the door at its peak, climbing up the outside ladder of the cylinder—all the way protected by an encased railing. When he reached the top he clambered up to the opening, grasping the railings tightly—as there was no encasement there—lest he fall to the concrete slab below. Every night after filling haylage into the silo, sometimes as late as two or four A.M., Larry went to the top to seal the upper door.

Being at the pinnacle of the blue silo made Larry feel that he had transcended the farming experiences of his grandfather and his father. They had always been close to the soil, close to the earth, and close to poverty; Larry wanted his family to rise up above that poverty. But the reality of his debts blindsided him. Silo salesmen wanted farmers to believe that they were not just "sodbusters" but that they were worthy of the respect accorded to agribusinessmen. Modern farmers were the equals of corporate mangers and those with M.B.A. degrees—salesmen could make it seem that being on the top of a blue silo was the same as standing on top of the IDS Tower in downtown Minneapolis. It almost seemed that Larry should be able to see that IDS Tower in the middle of a clear night. But on starry nights at the top of the silo, often long after midnight, he could barely see the dim lights of Redwood Falls, only twelve miles away. If he climbed the silo during the day, all he could see his was own farmstead with its alfalfa fields and watch the wind blowing through the green of the hay field, looking like the waves of the ocean. He had been entrusted with his family land, but now Larry felt adrift on those waves.

The haylage system was expensive but it had seemed his only recourse because the rest of us kids were leaving the farm behind; this system had promised to run an entire dairy farm on the labor of only two workers. By 1984, Chris was taking classes at Granite Falls Technical College, first enrolling in the auto-body repair program and then the fluid-power hydraulics program, but he came home to do the haylage harvesting for Larry that summer, living at Mom's house up on the hill. When there was not hay-work to do, then Chris did mechanic's work at our brother Jeff's garage in Morgan.

Larry had to use a farmer's wisdom to know how many acres of alfalfa to cut at a time. Thus, Larry looked at the current weather and at the weather report, and then had Chris cut as much as could likely be put up in the silo without being caught in any rain. They had to follow simple guidelines in order to harvest haylage under the proper timetable. If Chris cut the alfalfa in the evening, then they let the hay lay in windrows for forty-eight hours before chopping it. If Chris cut the hay in the morning and was done with the field by noon, then chopping began the next evening. The morning cutting was preferable: the cut and windrowed hay had less chance of getting rained on because it could be put up in the silo more quickly.[23]

Larry tried to work around the rainy days, but if precipitation fell on windrows of cut alfalfa, he had to let the windrows dry out again—without raking it to turn it over. He would never rake the alfalfa after it had been cut because the mechanical rake would knock too many leaves off the alfalfa stalks. Since haylage required a medium level of moisture, he only had to let the windrow dry again to the proper level and then chop it. The food value of the hay would deteriorate considerably, however, if another summer rainstorm doused the fields after a second drying-out period.[24]

Cutting hay with the swather was a straightforward operation. The swather moved rapidly at about five miles per hour through the deep-green alfalfa, putting the hay into orderly windrows, green against the clipped white stubble of the alfalfa stalks. The swather was set to cut each alfalfa plant at three inches above the ground. Larry taught Chris the two ways of knowing when the windrowed alfalfa had dried enough to be ready for chopping into haylage. Larry would touch with his fingers the green windrow in the hay field. If the alfalfa felt dry in the middle or bottom of the windrow, then it was ready.

The second method served to double-check the first test and was done while chopping the hay with the forage harvester. Chris would start chopping the hay, stop the chopper, and then take a handful of chopped hay from the forage wagon. He would squeeze the hay into the shape of a softball and then watch what happened when he quit compressing the hay. If the ball expanded and came slowly apart, then the alfalfa was at the right stage for chopping. If the hay was too moist, it would hold its shape as a ball and would not expand. If the hay was too moist, then Chris and Larry had to wait half a day or a whole day to allow further drying.

The chopping of the alfalfa took place after eight in the evening when the dew was full on the windrows. The dew helped retain the moisture in the hay and, in fact, gave back some of the moisture that the sun extracted during the day. Farmers treasured the dew in the haylage process, whereas farmers who made loose hay or baled hay disliked the dew because it combined with the air to cause heating, which damaged the hay during storage

Sometimes the dew did not appear until ten P.M., and then the chopping began. Chris was the one who usually drove the tractor that pulled the haylage chopper (otherwise called a forage harvester), with a silage wagon hooked on behind the chopper. Wendy occasionally drove the tractor for chopping when the boys could not help. She chose chopping because it was the easiest of the tasks involved with producing haylage. All the driver had to do was to drive the tractor around and around the alfalfa field, following the windrows of hay — it was like filling in the line on a connect-the-dots drawing. The biggest responsibility was to make sure to drive fast enough to chop the hay at the machine's maximum capacity but not so fast as to overload its *blowing* capacity, thereby clogging the chopper.

The chopper operator had to make sure that the windrow of cut alfalfa fed directly into the pick-up mechanism of the chopper, and the sharp knives inside the machine minced the alfalfa into tiny pieces. Larry would remove the knives and grind the blades to a razor's edge two or three times during the summer. He set the internal knives to a three-eighths-inch cut, so that the smaller pieces would be three-eighths of an inch in length, but some of the stems would be as long as an inch. (When cows ate the haylage the longer pieces were needed to provide roughage. The ideal haylage had twenty-five percent of the particles one and a half inches or longer, seventy-five percent of particles one and a half inches and shorter.) Because the hay was cut into small particles, it would pack well into the silo, thereby allowing few spaces for oxygen to be trapped where it could work on the hay's moisture to spoil the haylage.[25]

After the machine sliced the alfalfa, a fan propelled it into the forage-box wagon through a curved chute, producing a continuous flow from the cutter head to the blower fan to the spout. The chopping process released an aromatic fresh-hay smell, more extensive than experienced in any previous haying era due to the fact that the hay was more finely chopped and could release more fragrance.[26]

The main duty of the driver was to pivot the chute that propelled the chopped hay into the left, right, or middle of the haylage wagon behind the chopper. Adjusting the chute to fill the back corners and along the sides allowed for the maximum amount of haylage in the wagon. The driver looked back from the cab to see when the wagon was full, whereupon the driver would stop and wait for Larry to bring the next empty wagon. The only thing to do then was to ponder the light of the stars, or, perhaps, have the tune of "wait for the wagon" running through the driver's mind. It was easy to tell if it was overfull: haylage would fall out the front. The forage-box wagon consisted of two solid sidewalls, a back wall, and a half wall in the front that contained a mechanical unloader, all of which had a roof over it so the wind could not blow the haylage out.

The red International Harvester tractor had a cab, and a radio in it, but it was not air-conditioned. But in the dark, after ten, temperatures cooled off even in mid-July. Usually, the worst problem was staying awake. Larry drove the tractor that pulled the empty silage wagon to the field and hauled the fully loaded haylage wagon back to the blue silos for unloading. He or brother John, then seventeen, unloaded the haylage from the wagon by engaging the mechanical conveyor in the front of the haylage wagon, which unloaded the haylage into the feeding hopper where an auger brought it into a blower, which filled the blue silo. The forage blower, made by the Badger Equipment Company, propelled the haylage up to the top of the seventy-foot silo by means of the paddles on the blower fan, which revolved at a terrific speed. The rapidly revolving paddles worked as "throwing devices" to spew the haylage up the blower pipe and into the silo at a rate of from twenty to seventy tons per hour.[27]

Unloading was a dangerous job. The operator had to control two stationary tractors that powered two machines at the same time—the haylage wagon and the forage blower—and he had to stand in the midst of whirling PTO shafts, a moving auger and forage conveyor, and the blower's flywheel. He must be sure not to climb over any of the machinery because the surfaces could become slippery due to the polishing action over time of haylage abrading metal. He must keep his hands and feet away from the moving parts and take extreme caution while clearing blockages—making sure all the equipment was shut down.[28]

In July the sunlight lasted until ten at night, and the whole operation continued after dark by means of the tractor headlights. The trio of workers kept chopping haylage and pumping it into the silo until the day's cut was all gone. Depending upon the size of the alfalfa field and the amount of mechanical breakdowns, the chopping could go on until two or even later. One night, Larry worked all through the night and then attended to the cows' morning milking without a wink of sleep. He sent his brothers home to our mother's house to rest while he carried on.

The late-night work cast its spell on Chris once that summer. It was the darkest night of the year, in the dark of the moon and with clouds blocking out even the starlight. He drove the 806 tractor into the depths of the night, following the windrows of hay around and around the alfalfa field on "the hill." It was late, he was tired and sleepy-eyed from the flow of hot air from the engine to his face. At times the row of hay looked like the curb on a city street. When he had finished chopping the field, the last rows in the middle of the field, and had taken him in ever tighter and smaller squares. As he disengaged the chopper, he had lost track of which direction he needed to go to drive to the homeplace. He was lost, just a quarter of a mile away from the barn. He was so tired that he could not recognize which yardlights in the near distance belonged to which neighbor. He could not see the yardlight at Larry's place because the trees in the grove blocked it out. Chris kept driving in a circle in the night, looking for any sort of a landmark that could point him homeward. Lacking any rows of hay to guide him as to the points of the compass, he was trying to use the headlights of the tractor to give him a clue of the right direction. He looked for the ditchbank of the county ditch to the north or east of the field, but he could not find either of them. He could not find the gravel road to his west. At that point, Larry drove out to the field, driving the old WD-45 Allis tractor, which had no headlights. Chris could not even see Larry until Larry came into the range of Chris's headlights.

When both had come to a complete stop, Larry called to Chris: "What in the world are you doing?"

"I'm lost," Chris replied.

Larry told him, "Follow me."

And so Larry, no headlamps on his little Allis tractor, led the way back home. Both had experienced the disorienting effect of nighttime farming before; they understood how one could lose his sense of direction.

The haylage system was efficient, even if the timing of the work was exhausting. While the old-fashioned baling system required ten steps from field to feeding, the modern haylage method needed just four—cutting, chopping, unloading, and feeding. Each silo had an automatically operating unloader that dumped the haylage down to a conveyor that brought the hay to the cows in a feeding building. The haylage system reduced the amount of physical labor expended to get the hay under shelter and to the cows' mouths to the labor required to sit on a tractor seat and steer a machine.

Larry's methods defied conventional wisdom and traditional patterns of haying. He began cutting the alfalfa in late May, before any blossoms appeared, and kept cutting it every three weeks thereafter—just before the blooms were ready to appear—cutting the fifth crop in September. In this way he got five cuttings of alfalfa haylage in 1984. Traditionally, farmers had waited for the early blue, purple, and pink alfalfa blossoms to appear on the tops of the stalks before cutting the first crop in early to middle June. The conventional wisdom of agronomists held that alfalfa fields should not be cut between September 1 and the onset of the first killing frost in mid-October because the plants would suffer winter injury to a point beyond regeneration. Larry lost some of his alfalfa plants to winterkill because he had to get five cuttings of hay off his fields.

Larry managed to get away from his milking chores occasionally. He had taught John how to milk the cows when John was ten. When John was twelve he could be entrusted with all the milking chores—feeding the cows, milking them, cleaning up after milking—so that Larry and Wendy could get away for a day or several days for short vacations, which was something that our dad had never been able to do.

All summer, the three brothers cut and put up hay in their effort to fill the two blue silos. They filled the silos one at a time, working in evening and night hours. They fed the new haylage to the cows and kept refilling the silos as they unloaded them. They had instituted a new rhythm of haying on the land: swathing in the morning, chopping in the evening, filling the silo after suppertime. The haylage was of prime quality: green in color, rich in protein and feed value. The trio was able to get only enough haylage off of eighty acres of hay to feed the entire dairy herd, only enough corn from sixty acres to make ground feed, and enough oats from twenty acres of land to supplement the hay and corn. Larry had to plant five acres of land in sorghum as another forage crop to get more bulk feed, however. He made sure to cut and chop the grass along the road ditches because he needed the feed.

The alfalfa fields rebounded after each of the first four cuttings of hay, getting the rain in the usual amounts by means of summer showers and thunderstorms. The sun shone over the southwestern Minnesota prairie land as it had done for eons. The new pattern of haying made no difference to the land, for the earth there was so black and so rich that anything could grow on its face.

The new haylage system of haying created its own set of difficulties. As a mechanized system it was supposed to be a "hands-off" operation, and it was a laborsaving method except for the times when the machines did not work as well as they were supposed to. The haylage blower used for filling the silo occasionally plugged with hay, or something caught in the chopper, or the silage wagon broke down. These breakdowns forced Larry or Chris or John to stop the machine and clear the plug or to wield tools to repair chains or belts. Many times Larry or John had to climb up the ladder on one of the seventy-foot-high silos to unclog the blower pipe.

Chris had to clear excess alfalfa from the chopper knives to get it working smoothly. These eight wicked-looking knives usually chewed unmercifully any material put in their way, but even the chopper could be overwhelmed by a thick windrow of alfalfa stalks.

The expense of the haylage system was a constant background worry to Larry. A farm using haylage as feed for dairy cows was more expensive to operate than a dairy farm using baled hay. The blue silos were more expensive to build than a haybarn or a pole building or a tin shed; the swather was more expensive than a mowing machine; the chopper was more expensive than a baler; the silage wagons cost more than hayracks. Larry had to pay ten- to twelve-percent interest on his various loans as he carried a heavy debt load that totaled $250,000.

The monetary risk in farming in the 1980s was extremely high, but the modern haylage system also *heightened* risk of accidents rather than reducing them. Chopping hay and unloading silage wagons at night made fatigue and sleepiness unavoidable. The powerful chopper and the other machinery would be unforgiving to those who made a mistake or a misstep in the dark of night. No one was more aware of the dangers of farm machinery than Larry. By the early 1980s, Dad's death had become a cautionary tale, one of the most-remembered and retold PTO fatalities in Redwood County.

By 1984 Larry had replaced his first swather with a newer machine. The old swather, which had a twelve-foot-wide cutter head on
it, began to have problems cutting the alfalfa. It was a hydraulic problem, which in hindsight should have been easily fixed. But the ser-
viceman sent out by the implement dealer did not know his hydraulics very well, and the machine seemed to have become
unworkable. So Larry got rid of the old swather and bought a newer,
self-propelled swather.

The upgraded swather was huge and heavy, having its own engine and drive train. The sixteen-foot-wide cutter head was massive;
the cutter head itself weighted fifteen hundred pounds. The swather
featured an air-conditioned cab that kept out hay-dust and dirt and,
most importantly, the summer heat, thereby making hay cutting a
"no-sweat" proposition. With a nice stereo radio attached close to the
driver's chair, Chris could listen to country music on KLRG out of Redwood Falls, hearing singer Randy Travis and other new country
stars coming in clear as a bell on FM. It was the only station he could
get, but country music was the going thing for young people around
Morgan in those days.

The swather cut everything in its path in the summer of 1984,
from the first cutting of the hay in the week before Memorial Day to
the second cutting in the fourth week of June, the third cutting in the
third week of July, all the way to the fourth cutting in the middle of
August. Larry began cutting the fifth, and last, crop of alfalfa in late
September, while most other farmers were already done with their
third and last crop of alfalfa in August. Thus Larry's haying season
lasted about two months longer than that of other dairy farmers.

The extended haying season combined with late-evening to
middle-of-the-night chopping and silo filling took its toll on Larry.
That summer he was the most tired-looking soul around Morgan.
Even on Sundays, with relatives over for family gatherings, he would
be overcome by weariness, would lie down on his living-room carpet
and fall asleep instantly. Yet, day after day, he would rise early from
his bed at 5:00 A.M., like clockwork, and head down to the barn to
milk the cows. At 4:45 in the afternoon, like clockwork, he would go

to the barn for the evening milking. He was a dairy farmer, and that was his milking routine. It was good for the cows, even if it was exhausting for him.

He filled his minimal free time with activity. In the winter evenings of early 1984, he continued his habit of going downstairs to his woodshop to work on something made of oak. He didn't like watching television, he liked making something out of wood. Among the things he made was a little wall-shelf for me. He made it from oak—complete with beveled edges, all sanded, stained, varnished, and polished. He cut out the pieces for a wooden rocking horse, also of oak.[30]

He began to fix a beat-up 1949 Buick, having bought another old junker from South Dakota for parts. In the summer he worked to make the engine run again, no easy task after it had been sitting in a farm grove for thirty-five years. He was exhilarated after getting the motor to work, and he took Wendy for a drive in it to Mom's place on the hill. By the time they got back home he was so happy that he told her he had "a smile big enough for two people."[31]

1984 was a tremendous year for making haylage. The weather cooperated and rains came between cuttings and no rain fell on any of the hay in the field before it was chopped and blown into the blue silo. In mid-September he had to harvest his fifth crop of haylage from his alfalfa fields

Uncle LaVerne had noticed a change in Larry as August came to a close. Whereas Larry had always been worried about how much

Larry and his restorable 1940s Buick at Mom's place, May 1984. He had finally gotten the thing running.

hay he would get and how much he was carrying in debt, he appeared to be more at peace with his situation and more relaxed than LaVerne had ever seen him. The reality was, Larry knew that his debts had become too much for him to handle, that he had boxed himself in a corner and had no way to get out of it. He had come to realize that when the blue silo's lease period ended, in 1989, he would not have a credit rating that would warrant a banker loaning him money to purchase it. The silo company would take down the blue silo and he would lose his capacity for feeding his cows—and he would be bankrupt.

Larry could read the handwriting on the wall. His banker was already forcing him to farm in a manner that suited the banker but did not suit Larry. When Larry met with the banker, the man told him to begin raising his bull calves to become beef cattle rather than selling the young bull calves to other farmers. The banker said Larry needed this cash flow. Larry told him that he did not need the extra work involved in feeding the beef cattle and that the haylage he used to raise beef directly reduced the limited amount of feed that he had for his cows and young dairy heifers. The banker ultimately gave Larry no say in the matter.

The banker had also directed Larry to grow extra feed other than alfalfa for his livestock, in order to get more total tons of hay. The Harvestore professionals recommended quick-growing sorghum mixed with Sudan grass for additional feed in the blue silos, because it made such a large amount of feed. Larry questioned whether it was of sufficient quality to justify growing it, but he had to accede to this additional loan requirement in order to keep the banker happy.

All these things weighed heavily on Larry's mind as the summer of 1984 waned. By Labor Day, Chris had gone back for fall-quarter classes at Granite Falls; thereafter, he would make the thirty-mile drive only to come home for haying on the weekends. During the week, John was in classes for his senior year in Morgan, attending a newly named Cedar Mountain High School. Morgan and Franklin had paired their schools and renamed the high school after a hill between the two towns that was nothing like a mountain. By the last week of September, Larry had gotten just about all the alfalfa swathed, chopped, and blown into the blue silos, except for the last alfalfa field.

On the morning of September 29, a Saturday morning, Larry did the morning milking as usual. He awakened at five, ate just a roll washed down with milk, as was his custom, and went down to the barn and began the milking routine. He washed the udder of each cow with a washcloth and soap and water, attached the three milking machines to each of three cows, and let the pipeline carry the flow of milk to the bulk tank.

Larry listened to wcco Radio every morning on the barn's stereo system. He had already been up for an hour when announcers Charlie Boone and Roger Erikson sang their "Good Morning" song at precisely six o'clock.

Larry finished the milking and had fed the cows haylage in the outside feeder by nine, when he went into the house for breakfast with Wendy. The children were awake already and playing. Larry had asked John to come over in the morning to use the chopper to make some corn silage. Chris was coming over from Granite Falls and would arrive at noon to cut the alfalfa after Larry had greased the joints on the swather's head. The swather was parked beside the old chicken coop that had been converted to garage and storage space. It was right near the gasoline tank so that he could refuel it for a day's work. The swather was parked on a little raised portion of earth.

Wendy expected that Saturday to be like any haymaking Saturday, filled with work in the morning and into the late evening. She had allayed her fears of farmwork, having done some driving of tractors herself in her ten years of marriage to Larry. Yet she still worried about the dangers of farming. Because of the well-known story of Dad's accident, she was concerned about the rapidly whirling machinery and about general farm safety, as does every farm wife. She hated it when Larry would climb up to the top of the blue silos to check on the pipes. She hated that Larry would have to scurry up the silo ladder to the roof of the silo to close the top door of the blue silo every night after that day's haying was done. She worried because this always took place in the middle of the night. Wendy also worried whenever Larry went under a machine to repair it or grease it, whatever he had to do to it.

Larry went out to prepare the self-propelled swather that Saturday morning, and this time Wendy's worst fears were realized. The cutting head of the swather was extremely heavy, about fifteen hundred pounds of steel cutters and hydraulics. He crawled underneath the massive front of the cutter head to grease it, a task he had not often done before. He left the cutter head in the Up position and, according to the operator's manual, the mass of machinery was supposed to stay up while he was under it. He knew what the operator's manual said, but he did not know that when he bought the swather as used equipment, the dealer had given him the right manual for that model *but for the wrong year*. Normally he would put a block of wood under a machine like the cutter head to prevent it from coming down by accident, but there appeared to be little need of using the block this morning, for he would only be under it briefly. The machine was turned off and the cutter head was not supposed to come down unless the machine's engine was running.

But it did. He was lying on the highest part of the ground beneath the machine when fifteen hundred pounds of steel came down on his chest. The force pushed the air out of his lungs, and the cutter was too heavy to lift, too heavy even for his chest to expand enough to refill his lungs. He had drawn his last breath. After seven minutes without oxygen, his brain cells died and his body quit struggling to breathe. The weight of the machine broke no bones and left only faint marks on his chest. His body seemed unharmed.

John came near the swather to ask Larry where the other tractor was, so that he could get to work, but didn't see Larry. Supposing that he was in the house, he went there and talked to Wendy. She said that Larry was under the machine. They both rushed outside. Finding Larry there, John took a pry bar and lifted the whole weight off his brother, a task impossible without emergency-induced adrenaline.

John tried CPR, but failed to restart Larry's heartbeat or breathing. Wendy called for the Morgan ambulance, just as her mother-in-law had done sixteen years before. The rescue workers came and tried to resuscitate Larry. The crew got his strong heart to beat again and his lungs to work on the way to a hospital in the Twin Cities. But his time without oxygen had destroyed his brain cells. It fell upon Jeffrey to call the brothers who lived far away from Morgan. He telephoned Dana in Winona, where he worked as an electrician. He called me in Staples, where I worked as a history teacher. I remember the day with absolute clarity.

It was a Saturday. My wife Dianne was eight months pregnant; she had miscarried the year before. We were both worried that it could happen again. I was correcting papers at the dining-room table, listening to a record on the stereo. It was Mendelssohn. His Italian Symphony—romantic, melodic, lilting. The phone rang. Jeff was on the line.

"We have some trouble here," he said.

Larry had been taken to the University of Minnesota Hospital in the Cities. Jeff said it looked like Larry wouldn't make it. I just closed my eyes, knowing he wouldn't make it, but hoping he would. Why didn't we have *small* accidents, where he's hurt but will be okay? The Mendelssohn played on.

I tried to tell Dianne as calmly as I could; I didn't want to upset her in her eighth month. We prayed for Larry: what else could we do? I hoped for a miracle, but—just like Dad's—I knew there was no recovery from this accident. All that day we waited; that night, my cousin Michael called. He said the waiting was over.

Dianne and I wept together. Another call came, interrupting our grief, to deliver the same news. Hearing the message long distance was too much like coming home and hearing that Dad had died. I wasn't there. I never witnessed either death. I had to arrange for someone to take my classes on Monday. Dianne and I went to church the Sunday before leaving and were in the choir up front. I asked the congregation to pray for my brother's family. It broke my heart to have to stand there and tell everyone what had happened.

When we got to Morgan, Brenda needed to talk to me. We talked. John needed to talk to me, and we talked. I told him he had done all that anyone could, that he had done his best to save him. He told me his feelings, and I told him he had done well, and that he didn't have to take over the farm. He couldn't anyway.

The viewing of Larry's body took place in the same funeral parlor in Morgan where Dad's had taken place. Larry's hands were folded "peacefully" on his chest. Those hands that had worked so long and so well were the hands of a dairy farmer, not the hands of a carpenter. His hands looked the way his father's hands had looked: crooked, bent, misshapen, with deep cracks and calluses from exposure to the elements, and to water (from constantly washing the cows for milking twice a day, every day of the year). In coldest winter his hands had gotten chafed beyond healing. The cracked skin around the knuckles revealed the truth about dairying: Toil without end, amen.

Family and fellow church members of Bethany Lutheran gathered together for the funeral. The Danish neighbors who had mourned a father's passing, sixteen years later mourned a son's. The Basballes, the Christensens, the Hansens came to stand beside the Hoffbecks as they had all stood beside each other since emigrating from Denmark to the prairies of Redwood County. But the grief was compounded because this was the second farm accident on the same place, and the Hoffbeck farmstead was just a mile across the fields from the church, too close to be obscured. Even Pastor Vern Anderson was overcome with the loss, for he himself had come from a farming family near Belgrade, near Meire Grove in Stearns County, and he had cared for our family and other parishioners' families as they weathered the Farm Crisis of the 1980s. Anderson assured the mourners that the faith that carried Larry through his struggles had taken him "all the way to heaven."

The funeral was the most emotional I have ever attended. I will remember for the rest of my days the welling I felt in my chest when, on the way to the graveside, little Melissa sobbed, "I want my daddy." I knew how she felt. I wanted my brother back, my brother who had built the tree houses for us and had played softball on the lawn and had ridden bikes and shot cap guns with us, but this was not like the reenactment of Civil War battles of our childhood. When we counted to ten, brother Larry would not revive.

He died due to an innocent mistake. Had the ground underneath him at the accident site been only two inches lower, he could have continued to breathe and would have lived. Because of modern technology, Larry was officially alive when they arrived at the hospital; but as far as our family is concerned, he died under that swather.

His casket was made of oak.

Entries from My Mom's Farm Calendar, 1984

Saturday, September 29:
Larry died this eve. About 11:00 at the U of M in Minneapolis from a farm accident.

Monday, October 1:
Mary Larson and Linda Dahmes are doing the milking at Larry's. Larry Christensen is doing the feeding.

Tuesday, October 2:
So many decisions to make.

Wednesday, October 3:
Funeral for Larry at Bethany, 10:30, buried in church cemetery.

Thursday, October 4:
Got the beans combined at Wendy's and the hay made.

Epilogue

What pattern exists on these open prairies has been imposed on the land by farmers. They planted corn in straight rows, cut and raked their clover in windrows. When they graded gravel roads, they laid them out into a gridwork of one-mile squares. Those who baled their hay stacked it in patterns on the hayracks so that the greatest number of bales could be transported without sliding off. Haymakers tried to time the cutting of clover and timothy and alfalfa to dry spells in the weather systems. A schedule of crop rotation assures that alfalfa and soybeans put nitrogen back into the soil so that corn could take it out again. Order implied patterns, and patterns required order.

These patterns were followed by farmers but governed by weather and the land. They determined the tilling of the fields and the harvest of the hay, year after year, season to season, month by month. Most events on their farms were everyday and commonplace, so that daily chores were as natural to the farm landscape as are the trills of birds in the sky, the chirping of crickets under piles of straw, and the wind blowing through the elm trees in the grove. Andrew Peterson relied on the clockwork stroking of his scythe to reap a crop of wild grass. Oliver Perry Kysor heard the constant clicking rhythm of the horse-drawn mower as his faithful horses made their way through prairie grass and timothy grass. Gilbert Marthaler watched the recurrent beat of the workhorses' hooves on the ground as the hay loader pulled the hay from the ground in an unending circle of pulleys and chains and tines. Douglas Rongen heard a *click* every time the knotter turned out another bale of alfalfa hay. Larry and my other brothers made the tractors, choppers, and blowers chug and hum.

The dairy farm where I grew up followed a daily routine. We awakened early to begin the morning milking. The motions in milking—squeezing milk from the udder—were repetitive, rhythmical, by hand or by machine. Then we carried the milk to the milk room for storage in tanks or milk pails. We gave hay to the cows three times a day—morning, noon, and before nightfall. I was responsible for cleaning the manure out of the gutter every morning in winter, once a week in summer. These patterns provided a comforting stability to farm life. Sometimes they were all we had to rely on.

When Larry died he left behind a wife and three children. He had the foresight to carry a considerable amount of life insurance on himself, but it was not enough to pay off the farm debts and all other financial obligations. He knew full well that he might fall to a farm accident, as his father had and as his cousin Kelly had. Larry told Wendy to sell the cows if he died and then live on the farm by renting out the cropland.[1]

At first Wendy felt that she should not sell the cows or get out of the dairy operation that Larry had worked so hard to build. She felt that she would be betraying Larry and that selling the cows would be almost like selling a part of him. Since she had to make a living, she hoped just to keep the farm going.

Wendy advertised for a herdsman and hired a twenty-year-old man from Sleepy Eye who had taken a short course in dairy-farming herdsmanship at a technical school. He lived with the family, taking the spare bedroom in the basement. Sadly, his course of schooling had been too short. The inexperienced herdsman missed the vital breeding signs from the cows, and he failed to get them bred at the right time, losing months of the cows' natural lactation-milking schedule. He worked at the place for only four months, giving up because he found the work to be more than he could handle—for the farm had too much work for one person. Wendy tried to find a good long-term herdsman or herdswoman, but could not.

She found that trying to keep the farm and the dairy herd intact was too great a struggle. Hired help had proved to be of insufficient value and, at last, she decided to sell the cows in order to try to pay off the large blue silos, the farmland, and the house. The debts that she and Larry had carried together were too large for a person to hold alone. She first offered to sell the farm operation to Larry's brothers

John and Chris agonized about taking over the farm. Both knew that Larry had been "maxed out for what one person could handle" (as Wendy said),[2] and knew that he had done better than either one of them could do individually. Chris could have handled the fieldwork and the mechanical work, but did not like to milk cows. John enjoyed the dairy portion of farming but had no aptitude for the field operation or the mechanical end of things. Together they could have attempted it, but both knew that success was uncertain because they would have to pick up a new burden of debt in order to buy the farm from Wendy. She could not give them the farm because she had to provide for her three children.

Chris and Larry had talked about becoming partners prior to Larry's death, but he had already decided that agribusiness was not what he wanted to do for a living. He knew the workload and the debt load of modern farming, and these, combined with the twin tragedies at the farmstead, were hurdles too high for him.

John was at a crossroads in his life, but his path was clearly delineated, for he found that he could not bear to go to Wendy's farmplace anymore. He wanted to carry on the farming operation, but he was "afraid to farm the home operation or assist in any way," he said. The whole experience of our brother's death "proved to be an insurmountable emotional obstacle" for him at age seventeen, and he believed that he could not be "an asset to Wendy's attempt to continue the dairy operation." Wendy understood that John was not in a position to take on the farm alone at that point.

She sent a letter to each of Larry's father's brothers, giving them the opportunity to purchase the farmland and the homeplace. All of them said no. Only Uncle Vernie, who lived five miles to the south, seriously contemplated Wendy's offer to sell—his father George and his brother Raymond had farmed that land. Vernon knew in his soul that he should keep the farm in the family, but he let it go. As he said then, "It killed two."[3]

The patterns of death in rural communities are no less obvious than the patterns of life, if you look for them. In the natural cycle of things in Minnesota, death comes in the form of killing frosts and winter blizzards that follow. But ever since settlers began to work the prairie soil, the summer fields have held countless life-or-death moments. *Successful Farming* reported in 1985 that farmwork has been "the most hazardous occupation" of the twentieth century. More farmers are injured and more are killed in accidents than in any other industry in the United States. There are more deaths each year on America's farms than in its mines or quarries, more than in construction or factory work. At the time my brother was killed by the swather, the nation averaged 396 deaths per year due to farm accidents, more than double the second most dangerous job: operating lifting equipment like cranes and forklifts. Most tractor-related deaths during that time resulted from rollovers, like the one that claimed my cousin Kelly; between five and ten percent each year were caused by accidents, like my dad's, that involved a power-takeoff shaft. Two percent of the total machine-related deaths on farms came from entanglements with hay balers.[4]

Nevertheless, farmers continue to live out their days, tending their land, their crops, and their families. When tragedy strikes, the families are left to struggle without them, some reliving the events until they are consumed, others fleeing the painful memories. Most resign themselves to the rhythm of life and death on the farm and pray tragedy doesn't strike again. But in the 1980s—when debt ran through farming communities like a disorderly, eroding gully-washer—many farmers were forced to choose between making high-interest payments on expensive land and equipment or relying on old, outdated equipment that lacked current safety features.

By mid-decade, many farmers like my brother were both heavily in debt and reliant on unsafe equipment. When death struck these families, thus began another pattern that became the symbol for the Farm Crisis: the farm auction.

One of the young Danish-American neighbors, the second son of the Christensen's—Larry, L. C. he was called—bought Wendy and Larry's sixty-seven acres of farmland and the farmstead. Alvina retained ownership of the remaining ninety-nine acres and rented to L. C. After Wendy sold the farm property and had an auction to sell the farm equipment, she paid off her FHA loan on the land, the first blue silo, and the house.

L. C. Christensen milked cows there for a couple years, but dairying became increasingly difficult as the decade progressed. With milk prices going downward and feed prices going up, L. C. decided to quit milking cows. He tore the concrete floors and the cow stanchions out of the barn. He put in new concrete floors for conversion to a hog barn. He sold the milk pipeline system to another dairy farmer who wanted it. He painted the outside of the barn white; only dairy barns were red.

Larry's children, including Landon, took a last look at the haylage blower, baler, and bale elevator before the auction at the homeplace, 1986.

The silo company removed the leased silo from the property in the summer of 1986, marking the end of a dairy farm after three generations. Some observers referred to the remaining silos on the prairies as blue tombstones, and plenty of them stood silently on farms in the 1980s where the former owners had left, unable to make enough from the farm to pay their mighty debts.

On the day after Wendy got all of her furnishings and clothes—and all of Larry's woodworking tools and projects—out of the farmhouse, in February of 1987, she and Alvina went over "for a final look around" the homeplace. They looked in the outbuildings: the barn that had been their refuge in the winter storm of 1975, the garage where Larry had worked on that old car of Carl Jensen's, the new machine-shed where the swather had been kept, the old granary where the New Holland baler had always been parked. They walked the apple orchard where there had once been such a large harvest that Larry had to haul the gunnysacks of apples in the tractor and loader; through the grove where the best of Larry's tree houses had once overlooked the cowyard from its perch in the box elder tree; they saw the swing set that Larry had put up for Jessie, Melissa, and Landon to play on. Heartbroken, they walked the little rise near the granary, the place where Larry had silently died, just a hundred yards south of where Raymond died in the other accident. They looked at the new grove that Larry had planted, the ground snowy, the trees leafless. They did not say much, but they both knew that the farm was gone forever. They went into the house that Larry had built and sat together, in silence, on the couch that L. C. had bought along with the house. The two widows, not saying much, sat together on that couch in the empty house, until it got dark outside. There was no need to talk; they had been through the same experiences, both had the same feelings of loss. Wendy then went into town, to her father's house in Morgan. Wendy and the kids lived there for a short while until they moved to the Twin Cities, where she started a new life.[5]

Wendy later met widower Bob Hocking, whose wife had died of cancer, and they eventually married. He had one son named Tony.

Wendy's children, who had begun their lives as farmkids, became big-city kids. In the children's bedrooms in South St. Paul is a reminder of their father's handiwork: a combination desk, curio cabinet, and bookshelf unit, handmade from solid oak, stained and polished by their father in his woodshop.

Minnesota has witnessed the evolution of haymaking practices since Fort Snelling opened in the 1820s. Haycocks dotted the natural meadows as a part of making hay for the horses at the fort. By the 1860s, as white settlements and small farms extended along the Minnesota River, large haystacks stood in fields of timothy-grass stubble on both sides of the river near Mankato. By the 1880s, the era when railways brought new pioneers and haymaking machines to the Detroit Lakes region, sulky dump-rakes made long rows of clover hay near the railroad tracks. When lumberjacks marched out from their logging camps to cut down the white pine timber near Walker and Akeley in 1900, their sleighs moved past stables and nearby piles of hay that had been gathered from meadowlands near creek bottoms. In July of 1929 farmers near Albert Lea used side-delivery rakes to scoop leafy green alfalfa hay into long windrows that curved over the hills near the town. By 1959, bales lay strewn in hay fields near Hayfield, looking like enormous green Lego-blocks waiting to be packed on a pickup truck or hayrack. Today, Twin Cities suburbanites zoom "up north" on Highway 10 past Elk River, past Royalton, past Motley, to their lake cabins, the passage marked by large round bales of roadside grasses that lie in the ditches of the highways. They know less about cutting hay in a field of meadow grass in summer than they do about the physics of August heat shimmering over dark highways and city streets. Our closeness to the earth is fading as Minnesotans and as Americans.

My family is no exception. At the time of this writing, as we pass into a new century, none of the descendants of George Hoffbeck remain dairy farmers. Where all of his eight children had milked cows for a living, only seven of his forty-two grandchildren became farmers. By 1992, all seven had quit milking cows, though some turned to raising beef cattle. The changes in agriculture experienced during the 1980s were greater than the changes in Minnesota agriculture of any other decade. Thus, both in the immediate and extended families, dairy farming and the need to cut our own hay ended for the Hoffbecks.

from
The Last Mowing

There's a place called Far-away Meadow
We never shall mow in again,
Or such is the talk at the farmhouse:
The meadow is finished with men.

ROBERT FROST

Acknowledgments

I want to acknowledge the people who have helped me with researching and writing this unusual book about haymaking.

First, my deepest gratitude goes to my mentor, Arthur O. Lee, of Bemidji, for his wise advice, insights, and careful reading of the manuscript.

I am very grateful to those I interviewed, who gave of their time and memories, for this project would not be possible without them. My uncles Francis Hoffbeck, Vernon Hoffbeck, Norman Hoffbeck, and LaVerne Dahmes taught me much about hay and about my father. Gilbert Marthaler graciously told me of his experiences at Meire Grove. Doug Rongen, Art Rongen, and Marlys Rongen Lee were kind enough to share their personal papers and memories. Kenneth Kysor and Shirley Altoft mailed papers and photos from New York State. Mildred Basballe and Helen Nolting provided Danish perspectives; Wendell and Lois Setterberg gave insights into the Swedish past. Michael Dahmes, Tom Engel, Aunt Darlene Hoffbeck and Wayne Hoffbeck helped me with family lore. Earle and Mariann Dickinson assisted in so many ways, as did Tom and Liz Letson and Don and Suzanne Thomas.

I am deeply appreciative for the help given me at several museums and archives. I wish to thank especially John Decker at the Stearns County Museum, Leanne Brown in Carver County, Chris Schuelke and Kathy Evevold in Otter Tail County, Tamara Anderson-Edevold in Clearwater County, and Wanda Hoyum in Beltrami County. Sandy Slater found photos of haymakers in the University of North Dakota

Archives. Thanks also to the people at Agco Corporation and at Deere and Company for photo permissions.

At the Minnesota Historical Society, my gratitude flows to eight people who have been particularly important to me in this work. Jean Brookins and Greg Britton authorized the project. Anne Kaplan taught me about writing; Ann Regan nurtured the work all the way to the end; Debbie Miller provided the major research grant; editor Phil Freshman trimmed the many rough edges and honed the chapters in the first round of edits; and Will Powers tied the images to the text. I will always be grateful to editor Ted Genoways, truly a godsend, for his clear insights and his vision to put all the pieces together so wonderfully.

Thanks to Bill Mohn for illustrating haystacks and bales, and many thanks to the numerous individuals who shared their hayloft experiences and photos with me.

I am greatly indebted to my professors who taught me how to write history: Neil Stout for research methods and Samuel Hand for oral-history methods at the University of Vermont, and D. Jerome Tweton at the University of North Dakota. Even earlier, my teachers in Morgan—Barbara Valle, Jon Wittgraf, Paul Grogan, Rodney Harman, and Dorothy Lange—blessed me.

I sadly regret that Uncle LaVerne Dahmes and Uncle Vernie Hoffbeck passed away before this work was completed; they helped a fatherless boy grow up.

I give my personal thanks to friends who supported my efforts: Jim Musburger, Jim Koenig; Dan and Teresa Carlson; and Mark Junkert. In Minot, Brad and Anna Tengesdal, Greg and Karen Strand, Gary Ritchie, Norval Semchenko, and Howard Rogers. In Barnesville, Carlton and Barb Moe, Charlie Winkels, Norris Johnson, Howard and Linda Pender, Jeff Stangeland, Gary Nack, Jeff Berg, and Donald Swenson. From Morgan, Greg and Jeanne Green; Curt and Kathy Stokesbary; Mary Kay Weber Felty; Karen Koll Ingeman; Carl and Janet Basballe; Donna Christensen; and Kevin and Donna Kopischke. Others, too numerous to mention here, have helped me on my journey.

I am grateful for the assistance of my department chairs: Jonathan Wagner at Minot State University for German translations and counsel, and Paul Harris at Minnesota State University, Moorhead, for a number of illustrations. My colleague Ken Smemo gave key advice. Historian Steve Keillor lent encouragement.

And I express my deepest appreciation to my wife Dianne for contributing to this work in so many ways, through her love and devotion. And I happily thank my children, Leah, Katie, Mary, and Johnny for traveling with me to Phelps Mill and other places in Minnesota, making research more fun than it had a right to be.

And lastly, I thank God for my mom, who has loved me through all of the events herein inscribed and to this day.

Notes

Notes to the Prologue

1. William Shakespeare, *A Midsummer Night's Dream*, in *The Riverside Shakespeare* (Boston: Houghton Mifflin, 1974), 4.1.33.

2. Walt Whitman, *Specimen Days* (New York: Signet Classics, 1961), 152.

3. "What Does Your Memory Smell Like?" *USA Today* 120:2,560 (January 1992), 5; "Fond Memories of the Odor of Plastic," *Omni* 16:4 (January 1994), 33.

4. Helen Keller quoted in *Minneapolis Star-Tribune*, February 20, 1997, E1.

Notes to Chapter One

1. Quote from the King James Version of the Holy Bible, Jeremiah 29:11.

2. Birth of first three children: August 5, 1859; February 19, 1861; December 16, 1862. Andrew Peterson Diary, Emma M. Ahlquist's English translation. Andrew Peterson and Family Papers, 1854–1931, Minnesota Historical Society, St. Paul.

3. Josephine Mihelich, *Andrew Peterson and the Scandia Story* (Minneapolis: Ford Johnson Graphic and the author, 1984), 7, 8, 10, 11, 15, 16, 17. Carlton C. Qualey, "Diary of a Swedish Immigrant Horticulturist," *Minnesota History* 43 (Summer 1972): 64.

4. Solon J. Buck and Elizabeth Hawthorn Buck, *Stories of Early Minnesota* (New York: Macmillan, 1927), 180.

5. Peterson diary, August 30 and December 3, 1855; July 11, 1856; November 10, 1855.

6. Mihelich, *Andrew Peterson*, 8. Qualey, "Swedish Immigrant Horticulturist": 64. Grace Lee Nute, "The Diaries of a Swedish-American Farmer, Andrew Peterson," *Yearbook* (Minneapolis: American Institute of Swedish Arts, Literature and Science, 1945), 105.

7. Qualey, "Swedish Immigrant Horticulturist": 64. Hildegard Binder Johnson, "Factors Influencing the Distribution of the German Pioneer Population in Minnesota," *Agricultural History* 19 (January 1945): 50; in 1860, Waconia Township was seventy-six percent German and entirely encircled the Swedish community. Of the Germans: Peterson diary, August 10, 13, and 17, 1855.

8. Mihelich, *Andrew Peterson*, 40.

9. Ibid., 39.

10. Peterson diary, July 2, 1861.

11. Marilyn Arbor, *Tools and Trades of America's Past: The Mercer Collection* (Doylestown, PA.: Bucks County Historical Society, 1981), 48–49.

12. Peterson diary, July 16, 1862. Solon Robinson, *Facts for Farmers* (New York: A.J. Johnson, 1866), 35, 41.

13. Author's observation of Andrew Peterson farmstead, July 18, 1996.

14. Robinson, *Facts for Farmers*, 772.

15. Here and below, author interview of Isaac Palkki of Grand Rapids, MN, February 21, 1998, in Bemidji, MN; Palkki's Finnish-immigrant father was an expert with the scythe; also conversation with Jeff Stangeland, Barnesville, MN, November 21, 1999, regarding scythe handling.

16. "Swath," *Oxford English Dictionary* (Oxford: Clarendon Press, 1933), 290.

17. Alvin Jacobson, author interview, Deer Lake, MN, July 29, 1997. Notes in the author's possession. Now in his eighties, Alvin cut meadow grass along creek bottoms in northern Minnesota when he was young.

18. David Tresemer, *The Scythe Book* (Brattleboro, VT: By Hand and Foot, 1981), 24–25. Arbor, *Tools and Trades*, 49. Alvin Jacobson, interview.

19. The drying of leaves is described in New Idea, Co., *Modern Haying* (Coldwater, OH: New Idea, Inc., 1936), 13.

20. Peterson diary, July 23 and 25, 1862.

21. Alvin Jacobson, interview.

22. Peterson diary, July 24, 1862.

23. Ibid., July 24, 25, and 26, 1865.

24. Alvin Jacobson, interview.

25. The poles are described in Rodney C. Loehr, *Minnesota Farmers' Diaries* (St. Paul: Minnesota Historical Society, 1939), 15..

26. Marion K. Jameson, "Wright Rambles Then and Now," *Howard Lake Herald*, March 27, 1980, in "Agriculture" file, Wright County Historical Society, Buffalo, MN.

27. Pitchfork skills are described in Carl Werner, "Getting in the Hay," *Everybody's Magazine* 29 (July 1913): 109, 110.

28. Peterson diary, August 11, 1862.

Information about use of the feet comes from general knowledge of stack making.

29. Ormond H. Loomis, "Tradition and the Individual Farmer: A Study of Folk Agricultural Practices in Southern Central Indiana" (Ph.D. diss., Indiana University, 1980), 142, 143.

30. Topping a haystack, Peterson diary, August 25, 1857. Charles W. Dickerman, *How to Make the Farm Pay* (Philadelphia: Zeigler, McCurdy, c. 1869), 678.

31. Peterson diary, August 11, 1862. Information for this paragraph was gleaned from Peterson diary, September 1, 1863; September 2, 1864; July 26, 1865; and August 14, 1866.

32. Peterson diary, July 30, 1862. Fencing is explained in Merrill E. Jarchow, *The Earth Brought Forth: A History of Minnesota Agriculture to 1885* (St. Paul: Minnesota Historical Society Press, 1949), 7.

33. Jameson, "Wright Rambles."

34. Peterson diary, August 9 and 10, 1865.

35. Here and below, Kenneth Carley, *The Sioux Uprising of 1862*, 2d ed. (St. Paul: Minnesota Historical Society Press, 1976), 6–12. Also William Watts Folwell, *A History of Minnesota*, Vol. 2, (St. Paul: MHS Press, 1961), 109–146.

36. Philip O. Johnson, a Swedish settler in Carver County, as quoted in Mihelich, *Andrew Peterson*, 47.

37. Ibid.; and Peterson diary, August 20, 1862. General information from Carley, *Sioux Uprising*, 7–39. Peterson diary, August 24 and 25, 1862.

38. Carley, *Sioux Uprising*, 45, 48.

39. Peterson diary, September 5, 6, 7, and 8, 1862.

40. Carley, *Sioux Uprising*, 62ff.

41. Peterson diary, July 15, 1862; August 1, 1870.

42. Paul C. Johnson, *Farm Animals in the Making of America* (Des Moines, IA: Wallace Homestead Book Company, 1975), 62, 63. Ron Olson, author interview, Georgetown, MN, July 29, 1997. Notes in the author's possession.

43. Peterson diary, March 8, 1861. Charles W. Dickerman, *How to Make the Farm Pay* (Philadelphia: Zeigler, McCurdy, 1869), 392. Ron Olson, interview.

44. Johnson, *Farm Animals*, 62, 70; ox-goad, Peterson diary, April 23, 1860.

45. Wagon building, Peterson diary, September 30, 1861.

46. Harris P. Smith, *Farm Machinery and Equipment* (New York: McGraw-Hill, 1948), 286–87.

47. Peterson diary, November 18, 25, 1862.

48. Peterson diary, June 1, 1863; R. L. Allen, *Domestic Animals* (New York: Orange Judd, c. 1847), 192.

49. Mihelich, Andrew Peterson, 22. Hay for oxen, Robinson, *Facts for Farmers*, 778. Percy Wells Bidwell and John I. Falconer, *History of Agriculture in the Northern United States: 1620–1860* (New York: Peter Smith, 1941), 404–5. The going price for a pair of horses in 1851 was $100, while a pair of oxen cost $50; see Josiah T. Marshall, *The Farmers and Emigrants Hand-Book* (Hartford, CT: O. D. Case and Company, 1851), 16. Hiram M. Drache, *The Challenge of the Prairie* (Fargo, N.Dak.: Institute for Regional Studies, 1970), 140. Ron Olson, interview; Olson has trained both horses and oxen and has worked with oxen at various times since 1957. Dickerman, *How to Make the Farm Pay*, 385.

50. Ron Olson, interview. Also Thomas G. Fessenden, *The Complete Farmer and Rural Economist* (New York: A. O. Moore Agricultural Book Publisher, 1858), 64.

51. Peterson diary, January 19 and 27, 1865.

52. The amount is noted in Peterson diary, July 2, 1864. The wartime labor shortage is described in Jarchow, *The Earth Brought Forth*, 12, 15.

53. The fifty-percent figure is from Fort Snelling historic-site manager Stephen Osman, as quoted in "History Loud and Lively at Fort Snelling," *St. Paul Pioneer Press*, June 14, 1999.

54. Anderson in army, Peterson diary, August 27, 1864; and Mihelich, *Andrew Peterson*, 52. Neighbors Taylor August Johnson and Alfred Johnson also served in the Civil War: Mihelich, 41, 53.

55. Mower, Peterson diary, May 16, 1873; horses, February 22, 1877; horse-pulled hayrake, July 1, 1885. Jarchow, *The Earth Brought Forth*, 6, estimates the expenses of starting a typical 160-acre Minnesota farm in the 1860s at $795.

56. Timothy grass, Peterson diary, September 20, 1870; clover, July 9, 1872; alfalfa, May 17 and 21, 1886, and September 10, 1890. Timothy was commonly planted in southeastern Minnesota by 1862, clover in southern Minnesota from 1879 on. See Jarchow, *The Earth Brought Forth*, 239.

57. Barn foundation, Peterson diary, June 2 and 4, 1874; barn raising, June 25, 1874; hayloft, July 16, 1874; moving animals, December 17, 18, 19, and 22, 1874; second barn, May 20 and 29, 1884.

58. Grasshoppers, Peterson diary, August 31, 1876; buying hay, March 21, 1877; cutting graveyard hay, July 17, 1877.

59. Hail, Peterson diary, July 30, 1878; rainstorm, July 3, 1879; oppressive heat, July 12, 1880.

60. Andrew Peterson Farmstead, National Register of Historic Places Inventory Nomination form (Minnesota Historical Society, St. Paul, 1978), 3.

Qualey, "Swedish Immigrant Horticulturist": 69, 70.

61. Anna's death and burial, Peterson diary, September 22, 1889.

62. Here and below, Mihelich, *Andrew Peterson*, 142.

63. Qualey, "Swedish Immigrant Horticulturist": 63. Peterson Farmstead, national Register of Historic Places Inventory Nomination Form, 3.

Notes to Chapter Two

1. "I have been dreaming of home and Mother." O. P. Kysor Diary, Maine Township, MN, 1883 (Otter Tail County Historical Society, Fergus Falls, MN), January 28, 1883. Kenneth Kysor, *Time on My Hands* (Cattaraugus, NY: self-published, 1975), 22–24.

2. Drawn from the genealogy in Kenneth Kysor, "The Kysors of Cattaraugus County, New York" (typescript, Cattaraugus, N.Y.), 2.

3. Phone interview, June 5, 1997. Notes in the author's possession.

4. "The Northwestern Crop," *Fergus Falls Weekly Journal*, May 24, 1883.

5. Advertisement by Charles Wright, Fergus Falls real estate agent, *Fergus Falls Weekly Journal*, July 12, 1883.

6. "Homestead," *Fergus Falls Weekly Journal*, August 23, 1883.

7. Woods described in "Fergus Falls Next to Minneapolis," *Fergus Falls Weekly Journal*, October 4, 1883.

8. "The Park Region Recognized," *Fergus Falls Weekly Journal*, January 18, 1883.

9. "'The Young Man' Comes West," *Fergus Falls Weekly Journal*, January 4, 1883.

10. Minnesota State Census, 1885, for Maine Township, Otter Tail County (Otter Tail County Historical Society, Fergus Falls). Fifty-one of the 300 peo-

ple then living there had come from New York.

11. Norwegian Grove in "Otter Tail Towns: Where They Got Their Names," *Fergus Falls Weekly Journal*, March 15, 1883. June Drenning Holmquist, ed., *They Chose Minnesota: A Survey of the State's Ethnic Groups* (St. Paul: Minnesota Historical Society Press, 1981), 59–61.

12. *Fergus Falls, 1872–1972* (Fergus Falls: Centennial Corporation, 1972), 2, 3, 8, 10, 13, 14, 20, 21. Ralph W. Hidy, Muriel E. Hidy, and Roy V. Scott with Don L. Hofsommer, *The Great Northern Railway: A History* (Boston: Harvard Business School Press, 1988), 22, 318.

13. "City of Fergus Falls," *Fergus Falls Weekly Journal*, December 27, 1883; "Fergus Falls Next to Minneapolis," idem, October 4, 1883. Chicago tourists in "Life at Fish Lake," idem, September 6, 1883. Bears in "Bear Hunt in Maine," idem, September 20, 1883. "Summering at the Lake," idem, August 9, 1883.

14. Phone interview, June 5, 1997. Notes in the author's possession.

15. Kenneth Kysor, *Time on My Hands*, 24.

16. Kysor diary, September 4, 7, 12, 17, 24, 27, and 28; October 3, 4, 10, 12, and 18, 1883.

17. Another local farm listed a log stable twenty-five by forty-five feet in "Farm!" (advertisement), *Fergus Falls Weekly Journal*, July 5, 1883. Kysor diary, March 5 and 7, 1883.

18. Samuel N. Stokes, *Saving America's Countryside: A Guide to Rural Conservation* (Baltimore: Johns Hopkins University Press, 1989), 40.

19. Charles D. Babcock, "Kysor Genealogy Compiled by Charles D. Babcock" (manuscript, 1968), 48, 49. Copy in the author's possession.

20. Secoy barn raising in "Maine,"

Fergus Falls Weekly Journal, July 26, 1883; and Kysor diary, July 25, 1883.

21. Kysor diary, April 6, 1883. Kysor does not list the dimensions of his barn but the estimated size comes from typical barns of the time, which are described in Byron D. Halstead, *Barns, Sheds and Outbuildings* (Brattleboro, Vt.: Stephen Greene Press, 1977), 29–36; this is a reprint of a classic nineteenth-century farm-plan book. Three-bay barns are explained in Allen G. Noble, *Wood, Brick, and Stone*, vol. 2 (Amherst: University of Massachusetts Press, 1984), 16–18.

22. Solon Robinson, *Facts for Farmers* (New York: A.J. Johnson, 1866), 584.

23. Apple trees, Kysor diary, April 30 and May 8, 1883.

24. Popcorn, Kysor diary, March 27, 1883; cold, January 15, 1883. Yankee customs are noted in June Drenning Holmquist, ed., *They Chose Minnesota: A Survey of the State's Ethnic Groups* (St. Paul: Minnesota Historical Society Press, 1981), 64. Polar waves in *Fergus Falls Weekly Journal*, January 18, 1883; Robinson, Facts for Farmers, 393.

25. Punishment, Kysor diary, June 6, 1883; cider, June 2, 1883. Ages from Bureau of the Census, *U.S. Census, 1870* (New York, Cattaraugus County, Leon Township), 4, 27.

26. "Maine," *Fergus Falls Weekly Journal*, December 8, 1910. Bureau of the Census, *U.S. Census, 1870*, New York, Cattaraugus County, Leon Township, 27; Bureau of the Census, *U.S. Census, 1880*, New York, Cattaraugus County, Leon Township, 4. Visit, Kysor diary, November 3, 6, and 18, 1883. Ages from Bureau of the Census, *U.S. Census, 1880* (New York, Cattaraugus County, Leon Township), 4.

27. "Preaching in Maine," *Fergus Falls Weekly Journal*, September 20, 1883;

"Divine Service In Maine," idem, September 6, 1883; "Preaching in Maine," idem, August 23, 1883. Locations listed, Kysor diary, March 30; May 11; March 11, April 29, May 10, May 27, and May 13, 1883.

28. Singing, Kysor diary, March 27, 1883; taffy, March 31, 1883. Play in "The Drama in Town of Maine," *Fergus Falls Weekly Journal*, March 15, 1883; and Kysor diary, March 20, 1883.

29. "State of Maine Natives Founded Township in 1871," *Fergus Falls Daily Journal*, July 25, 1968, in Maine Township file (Otter Tail County Historical Society, Fergus Falls, MN). Berries, Kysor diary, August 1 and 6, 1883.

30. "Death of a Pioneer Woman," *Fergus Falls Weekly Journal*, January 25, 1883.

31. Kysor diary, June 20 and 23, 1883.

32. Bill Buckingham, "The Agricultural History of Otter Tail County," Agriculture file, Otter Tail County Historical Society, Fergus Falls, MN

33. Chores, Kysor diary, January 7, 1883; blizzard, idem, January 9, 1883.

34. Sunstroke, Kysor diary, July 3, 1883; rain, July 19, 1883; bugs, July 4, 1883. Robinson, *Facts for Farmers*, 439.

35. Grubbing, Kysor diary, April 17, 1883; wheat and oats, April 12 and following; peas, April 30; potatoes, May 1; corn, May 16, June 8; sweet corn, May 16; cabbage, May 25; beans, June 18.

36. E. P. Powell, "In the Hayfield," *Outing Magazine* 52 (June 9, 1908): 349.

37. Kysor diary, in passim, 1883.

38. "Our Western Town," *Fergus Falls Weekly Journal*, October 4, 1883.

39. Robinson, *Facts for Farmers*, 753. Shoots discussed, Arthur O. Lee (historian), interview, Bemidji, MN, June 24, 1996; notes in the author's possession.

40. Robinson, *Facts for Farmers*, 750.

41. Harris P. Smith, *Farm Machinery and Equipment* (New York: McGraw-Hill, 1948), 266–267.

42. U.S. Department of Agriculture, *Grass: The Yearbook of Agriculture, 1948* (Washington, D.C.: Government Printing Office, 1948), 174.

43. Carl Werner, "Getting in the Hay," *Everybody's Magazine* 29 (July 1913): 106. Clockwise movement discussed in *Modern Haying* (Coldwater, OH: New Idea, 1936), 19.

44. Walter A. Dyer, "Chronicles of a Countryman," *Country Life* 52 (July 1927): 70.

45. Arthur O. Lee, interview.

46. Dyer, "Chronicles of a Countryman," *Country Life* 52 (July 1927): 70.

47. Robinson, *Facts for Farmers*, 774.

48. Haycocks, Kysor diary, July 30, 1883. Charles W. Dickerman, *How to Make the Farm Pay* (Philadelphia: Zeigler, McCurdy, 1869), 760. Timothy in U.S. Department of Agriculture, *Grass*, 684–685.

49. Powell, "In the Hayfield," 348.

50. Kysor diary, July 31, 1883.

51. Bickford, Kysor diary, August 13, 1883. Thomas G. Fessenden, *The Complete Farmer and Rural Economist* (New York: A. O. Moore Agricultural Book Publishing, 1858), 276.

52. Food, Kysor diary, August 22, 1883. Ridge pole/tent, Kysor diary, April 9, 1883. Robinson, *Facts for Farmers*, 774. Charles W. Dickerman, *How to Make the Farm Pay* (Philadelphia: Zeigler, McCurdy, 1869), 679.

53. George's injury, Kysor diary, August 22, 29, and 30, 1883. "Be Careful," *Fergus Falls Weekly Journal*, August 16, 1883.

54. Werner, "Getting in the Hay," 106. Robinson, *Facts for Farmers*, 298; idem, "cool hands and face," 982.

55. Prairie work, Kysor diary, August 22, 1883.

56. Picnic and more work, Kysor diary, September 5, 1883.

57. Recipe for switchel in David Tresemer, *The Scythe Book* (Brattleboro, VT: By Hand and Foot, 1981), 63.

58. "Summer Drinks," *The Farmer*, July 15, 1905, 471.

59. Hungarian grass, Kysor diary, September 13, 1882. U.S. Department of Agriculture, *Grass*, 691–692. Kysor diary, September 13 and 14, 1883.

60. Robinson, *Facts for Farmers*, 774–775.

61. "Girard," *Fergus Falls Weekly Journal*, July 19, 1883.

62. Gilbert Rogers, "Reminiscences" (manuscript, Goodhue County Historical Society, Red Wing, MN, 1978), 1.

63. Earle Dickinson, interview, Bemidji, MN, March 19, 1997. Notes in the author's possession. Dickinson made loose hay in his youth and baled hay in his later years.

64. Information about hayracks from *Franklin Steele, Jr., and Company's Illustrated Catalog of Agricultural Implements* (Minneapolis: Johnson, Smith and Harrison Printers, c. 1882).

65. Hayrack, Kysor diary, December 6, 1883; seven loads, November 23, December 2, 4, and 6, 1883.

66. Lightning-rod salesman's visit in "Maine," *Fergus Falls Weekly Journal*, July 26, 1883.

67. Tom Simkin and Richard S. Fiske, *Krakatau 1883: The Volcanic Eruption and Its Effects* (Washington, D.C.: Smithsonian Institution Press, 1983), 15, 49. The sun-glows were first reported in North America at Yuma, Arizona, on October 19 and were noted in the eastern United States on October 30, 1883. Specific mention of the sunsets in editorial, *Yellowstone Journal* (Miles City, Mont.), December 22, 1883; and in *Fergus Falls Weekly Journal*, December 27, 1883. Eruption in

"Worst in a Century," *Fergus Falls Weekly Journal,* August 30, 1883.

68. Rosanne Begantine Giencke, *Phelps: A Peek into its Past* (Fergus Falls, MN: Rosanne Giencke, 1911), 2, 13, 19. 1885 *Minnesota State Census,* Maine Township, Otter Tail County, 13.

69. "Battle Lakelets," *Battle Lake Review,* July 28, 1887; also "Cemetery Records, Otter Tail County, MN, Maine Township, Silent Vale Cemetery," vol. 5 (Otter Tail County Historical Society, Fergus Falls, MN), roll 36. Kysor, "Kysors of Cattaraugus County," 27.

70. Giencke, *Phelps,* 2, 19. 1885 *Minnesota State Census,* Maine Township, Otter Tail County, 16.

71. Obituary, Caroline (Mosher) Kysor in "Maine," *Fergus Falls Weekly Journal,* December 8, 1910.

72. "Death of an Old Maine Resident," *Wheelock's Weekly* (Fergus Falls, MN), August 8, 1912.

73. Caroline Kysor entry in Kysor diary, November 1, 1883.

74. "Cemetery Records, Otter Tail County, MN, Maine Township, Silent Vale Cemetery," vol. 5 (Otter Tail County Historical Society, Fergus Fall, MN), roll 36.

75. Giencke, *Phelps,* 8.

Notes to Chapter Three

1. Interview with Gilbert Marthaler, Meire Grove, MN, by John Decker, January 25, 1977, tape #1349 (Stearns County Historical Society, South St. Cloud, MN), transcript p. 6. I chose 1924 because Marthaler recalled the church burning down around that time.

2. Interview with Gilbert Marthaler, Meire Grove, MN, by the author, July 22, 1996, 7; notes in the author's possession. Unless indicated otherwise, all information about the Marthaler family comes from this interview.

3. Here and below, see "Meire Grove," *St. Cloud Daily Times,* July 27, 1899, 2; Paulin Blecker, *Deep Roots: One Hundred Years in Meire Grove* (St. Cloud: Sentinel Publishing Co., 1958), 61; "Victim of Lightning," *St. Cloud Daily Times,* July 26, 1899, 3; "Killed by Lightning," *St. Cloud Daily Times,* July 27, 1899, 4.

4. Patricia Marthaler Miller, *Marthaler Genealogy* (Sauk Centre, MN: self-published, 1979), 2 and preface.

5. Ibid., 3, 4.

6. Ibid., 12–13.

7. Bob Artley, *A Book of Chores, As Remembered by a Former Kid* (Ames, IA: Iowa State University Press, 1989), 28, 29.

8. Fordson information in a letter from Gilbert Marthaler to the author, August 15, 1996, 1. Allen R. Yale Jr., *While the Sun Shines: Making Hay in Vermont 1789–1990* (Montpelier, VT: Vermont Historical Society, 1991), 33.

9. Shade details added from a letter from Harlan Dalager, Minnetonka, MN, December 18, 1996, to the author, 1; Dalager grew up on a farm in western Minnesota. Sharp blades idea in Harris P. Smith, *Farm Machinery and Equipment* (New York: McGraw-Hill, 1948), 274.

10. Interview with Anton Stern, Belgrade, MN, by the author, July 22, 1996, 5; notes in the author's possession.

11. Clover and alfalfa are legumes, which add nitrogen to the soil. Fields in which row crops, such as corn, were grown a few years consecutively lost nitrogen; clover and alfalfa restored this precious element to the soil by converting it from the air through its roots. University of Minnesota agricultural experts had promoted the planting of legumes after the turn of the century. Paul C. Johnson, *Farm Inventions in the Making of America* (Des Moines, IA:

Wallace-Homestead Book Company, 1976), 97.

12. W. R. Humphries and R. B. Gray, *Partial History of Haying Equipment,* (Beltsville, MD: U.S. Department of Agriculture, 1949), Information Series, 74:25.

13. Ibid., 74:24.

14. Ibid., 74:31–36; Sears, Roebuck & Company sold a hay loader for $45.00 in 1900 and for $31.20 in 1902; see *Sears, Roebuck and Company Consumers Guide, Fall 1900* (Northfield, Ill.: DBI Books, Inc., reprint edition, 1970), 944; and *The 1902 Edition of the Sears, Roebuck Catalog* (New York: Bounty Books, 1969), 687—the price of a hay loader in 1902 was roughly equal to that of a farm wagon, iben, 378. The 1892 International Harvester–McCormick catalog pictured machines as the way to "agricultural prosperity," see drawing in "McCormick–International Harvester Company Collection," State Historical Society of Wisconsin Archives website http://www.shsw.wisc.edu/.

15. Letter from Harlan Dalager, Minnetonka, MN, December 18, 1996, to the author, 1. Johnson, *Farm Inventions in the Making of America,* 104–105.

16. Artley, *A Book of Chores,* 28, 29.

17. Hay-slings, Dalager letter, December 18, 1996, 1.

18. Conversation in St. Cloud with Maynard Sand, of St. Wendel, MN, October 8, 1999; notes in the author's possession.

19. Size of the barn, interview with Gilbert Marthaler by John Decker, tape #1349, transcript p. 18.

20. Gilbert Marthaler, author interview; with additional details added from a letter from Harlan Dalager, to the author, 1. Hay carrier in Edward H. Knight, *Knight's New Mechanical Dictio-* *nary* (New York: Houghton, Mifflin & Co., 1884), 448.

21. Letter from Harlan Dalager to the author, 1.

22. Interview with Gilbert Marthaler by John Decker, tape #1349, transcript p. 21.

23. Ibid., 8, 21.

24. "Gored to Death by Bull," *Melrose* (MN) *Beacon,* March 13, 1924, 1.

25. "Vicious Bull Causes Death of Villard Man," *Melrose Beacon,* June 12, 1924, 1.

26. "Henry Kruse Injured by an Enraged Bull," *Melrose Beacon,* November 20, 1924, 1.

27. Telephone conversation with Gilbert Marthaler, Meire Grove, MN, June 11, 1997; notes in the author's possession. "Tornado Strikes Near Greenwald," *Melrose Beacon,* July 24, 1924, 1.

28. "Lightning Causes Large Barn Fire," *Melrose Beacon,* September 4, 1924, 1. "Barn Burns," *Melrose Beacon,* September 25, 1924, 1.

29. "Two Deaths Here During Past Week," *Melrose Beacon,* September 11, 1924, 1.

30. "Prominent Meire Grove Lady Dies," *Melrose Beacon,* December 18, 1924, 1.

31. "Large Attendance at Church Dedication," *Melrose Beacon,* October 9, 1924, 1; "Meire Grove Will Hold Dedication Celebration," *Melrose Beacon,* September 25, 1924, 1.

32. "Region Started As Producer of Big Grain Crops," *St. Cloud* (MN) *Daily Times,* July 17, 1931, 15.

33. "Butter Firm at Meire Grove Is Set Up in 1897," *St. Cloud Daily Times,* June 21, 1943, 18.

34. "Minnesota 13—It Turned Corn Into Gold," *St. Paul Dispatch,* October 12, 1962, 1, Section 2, in "Prohibition"

file, Stearns County Historical Society, St. Cloud, MN

35. Interview with Gilbert Marthaler by John Decker, transcript p. 16.

36. *Stearns County Directory, 1910* (St. Cloud: Birney Moore, 1910), 5, 23, 86, 88, 97. *St. Cloud City Directory, 1910* (St. Cloud: John H. Ley, 1910), 247.

37. Interview with Michael C. Schneider, St. Cloud, MN, by Mildred M. Dumonceaux (Stearns County Historical Society, St. Cloud), tape #1352, transcript p. 14.

38. "Holdingford stayed 'wet' during dry spell," *St. Cloud Times*, March 29, 1996, 1(B).

39. Interview with Gilbert Marthaler by John Decker, transcript p. 17.

40. Interview with Michael C. Schneider, transcript p. 14.

41. Interview with Gilbert Marthaler by the author; notes in the author's possession.

42. Miller, *Marthaler Genealogy*, 4.

43. U.S. Department of Commerce, Bureau of the Census, *1992 Census of Agriculture*, vol. 2, part 3, "Ranking of States and Counties," (Washington, D.C.: GPO, 1994), 32, 86.

44. Anthony Rosycki, "The Evolution of the Hamlets of Stearns County, Minnesota" (M.A. thesis: University of Minnesota, 1976), 2, 23, 48.

Notes to Chapter Four

1. Dorothy Burton Skardal, *The Divided Heart: Scandinavian Immigrant Experience Through Literary Sources* (Lincoln: University of Nebraska Press, 1974), 105.

2. Garfield: *The First 100 Years, 1880–1980* (Fertile, MN: Garfield Centennial Committee, 1980), 12, 13.

3. Ibid., 7, 45.

4. Ibid., 4.

5. Interview with Doug Rongen,

Fertile, MN, by the author, July 25, 1996, notes in the author's possession.

6. Garfield: *The First 100 Years*, 31.

7. Marlys (Rongen) Lee, "Art Rongen's Life" (typescript, 1995), 4; copy in the author's possession.

8. *Garfield: The First 100 Years*, 25. Silo in Lee typescript, 4.

9. Interview with Marlys (Rongen) Lee, Mayville, ND, October 5, 1996, in the home of her son, Cal Lee, Glenburn, ND; notes in the author's possession.

10. Letter from Doug Rongen, Fertile, MN, April 15, 1997, to the author, 16.

11. Letter from Marlys (Rongen) Lee, Mayville, ND, March 24, 1997, 2.

12. Marlys (Rongen) Lee, "Childhood Memories of Charles for His Children by Their Rongen Uncles and Aunts" (typescript, 1990), 7, copy in the author's possession.

13. Interview with Marlys (Rongen) Lee. Letter from Marlys (Rongen) Lee to the author, 2, 3.

14. Interview with Marlys (Rongen) Lee.

15. Lee, "Childhood Memories," 1, 7.

16. Letter from Doug Rongen to the author, 10.

17. Interview with Marlys (Rongen) Lee.

18. Ibid.

19. Letter from Marlys (Rongen) Lee to the author, 8. Lars K. Fadness was born April 26, 1868, and died January 4, 1891, according to the tombstone in the Little Norway Lutheran Church cemetery.

20. Ibid., 4, 9.

21. Ibid., 5.

22. Ibid., 7.

23. Letter from Doug Rongen to the author, 8. "Last Rites Held July 25 For

Lars J. Rongen," *Fertile Journal,* July 29, 1959, 1.

24. Ibid. "Lars J. Rongen Rites Held Saturday," *Fertile Journal,* July 29, 1959, 3.

25. Letter from Doug Rongen to the author, 5. Interview with Marlys (Rongen) Lee.

26. Letter from Doug Rongen to the author, 9.

27. Interview with Doug Rongen.

28. Herbert W. Congdon, *The Covered Bridge: An Old American Landmark* (Middlebury: Vermont Books), 82.

29. Advertisement for Case balers, *The Farmer* (July 3, 1943), 10; W. R. Humphries and R. B. Gray, *Partial History of Haying Equipment,* U.S.D.A. Information Series no. 74 (Beltsville, MD: U.S. Department of Agriculture, 1949), 54.

29. Wheeler McMillen, "The Ancient Technology of Farming: Ohio, 1910," *American Heritage of Invention & Technology* 7, no. 1, (Spring/Summer 1991), 49.

30. David B. Danbom, *Born in the Country: A History of Rural America* (Baltimore: Johns Hopkins University Press, 1995), 233, 234, 240.

31. Letter from Doug Rongen to the author, 6.

32. Conversation of Mary Kay (Weber) Felty, daughter of Clarence Weber, Morgan, MN, with the author, July 27, 1996; notes in the author's possession.

33. Letter from Doug Rongen to the author, 5, 10.

34. Conversation with Earle Dickinson, Bemidji, MN, July 21, 1996; notes in the author's possession.

35. Conversation with Arthur O. Lee, Bemidji, MN, June 24, 1996; notes in the author's possession.

Notes to Chapter Five

1. Basballe in "Viet Nam G.I. Will Return for Second Hitch," *Morgan* (MN) *Messenger,* January 11, 1968, 1; Kerkhoff in "I'm Scared But Not Chicken," *Morgan Messenger,* June 6, 1968, 1.

2. "Real Estate Taxes are Due Oct. 31st; Board Seeks Disaster Area Classification," *Morgan Messenger,* October 31, 1968, 1.

3. "P.T.O. Cause of Fatality," *Morgan Messenger,* November 21, 1968, 1; "Rites Are Conducted for Victim of Farm Accident on Friday," *Morgan Messenger,* November 28, 1968, 1.

4. Charles W. Howe, *Forty Wonderful Years: Morgan, Minnesota and Vicinity, 1876–1916* (Morgan: Morgan Messenger, 1916), 17.

5. "Many Giving Up Their Herds," *Morgan Messenger,* July 4, 1968, 8.

6. U.S.D.A. statistics cited in "Agricultural Products Provide Top industry in State of Minnesota," *Morgan Messenger,* July 4, 1968, 6.

7. Phone interview with Wendy Sandmann Hoffbeck Hocking, St. Paul, MN, January 18, 1998; notes in the author's possession.

8. Letter from Alvina Hoffbeck, Morgan, MN, to the author, July 13, 1976, 2. Fifty-six bales, in letter from Alvina Hoffbeck, Morgan, MN, to the author, June 10, 1975, 1.

9. University of Alberta website, 1998, "Feed Storage," http://www.afns.ualberta.ca

10. Minnesota Historical Society, *Minnesota Farmscape: Looking at Change* (St. Paul: Minnesota Historical Society Press, 1980), 10.

11. Letter from Alvina Hoffbeck, Morgan, MN, to the author, June 27, 1974, 2.

12. "Sealed Storage: Will it fit your livestock system?" *Successful Farming* 69 (June/July 1971), B1, B5.

13. Sale of bale machine in letter from Alvina Hoffbeck, Morgan, MN, to the author, July 20, 1976, 2.

14. "Sacred Cow," *Time* (February 18, 1985), 26.

15. Letter from Wendy Sandmann Hoffbeck, Morgan, MN, to the author, August 2, 1977, 2.

16. Warranty deeds, Redwood County, Alvina Hoffbeck to Larry & Wendy Hoffbeck, Vol. 134, December 12, 1977, p. 359. Mortgages, Redwood County, Vol. 169, FmHA to Larry & Wendy Hoffbeck, December 12, 1977, p. 251. Harold D. Guither and Harold G. Halcrow, *The American Farm Crisis: An Annotated Bibliography with Analytical Introductions* (Ann Arbor, MI: Pierian Press, 1988), 145.

17. Twine price in letter from Alvina Hoffbeck to the author, June 27, 1974, 2. Letter from Alvina Hoffbeck to the author, July 20, 1976, 2.

18. Letter from Alvina Hoffbeck, Morgan, MN, to the author, October 27, 1976, 3.

19. "Randolph County, Illinois: Harder Work, Less Cash, More Debt," *Successful Farming* (January 1984), 21.

20. Jeffrey Burke, "Country Wisdom," *Harper's Magazine* (July 1980), 81; William Mueller, "Growing with the Times," *Harper's* (July 1980), 84.

21. "America's Farmers Down the Tubes?" *U.S. News & World Report* (February 4, 1985), 47.

22. Recession in Ed Magnuson, "Real Trouble on the Farm," *Time* (February 18, 1985), 27.

23. John L. Hoffbeck, "Haymaking as Part of My Farming Experience," August 20, 1995, letter to the author.

24. Conversation with Chris Hoffbeck, Grand Rapids, MI, February 1, 1998; notes in the author's possession.

25. "How to Make Perfect Haylage," *Successful Farming* (June/July 1983), D7. Allen R. Rider and Stephen D. Barr, *Hay and Forage Harvesting* (Moline, IL: Deere & Company, 1976), 249.

26. Letter from Chris Hoffbeck, Grand Rapids, MI, February 2, 1998, 1.

27. Rider and Barr, *Hay and Forage Harvesting*, 288, 289.

28. Ibid., 300–301.

29. University of Minnesota agronomist Oliver Strand quoted in "Alfalfa Needs Growth in Fall," *Redwood Gazette*, August 17, 1967, 4.

30. Letter from Wendy Sandmann Hoffbeck, Morgan, MN, to the author, February 1984, 1.

31. Phone interview with Wendy Sandmann Hoffbeck Hocking.

Notes to the Epilogue

1. Phone interview with Wendy Sandmann Hoffbeck Hocking, St. Paul, MN, January 18, 1998; notes in the author's possession.

2. Ibid.

3. Conversation with Alvina Hoffbeck, Morgan, MN, October 17, 1996; notes in the author's possession.

4. Cheryl Tevis, "Farming: America's Most Hazardous Occupation," *Successful Farming* 83 (April 1985): 16–17.

5. Letter from Alvina Hoffbeck, Morgan, MN, to the author, February 26, 1987, 2.

Bibliography

Books

Allen, R.L. *Domestic Animals.* New York: Orange Judd & Company, 1847.

Arbor, Marilyn. *Tools & Trades of America's Past: The Mercer Collection.* Doylestown, PA: Bucks County Historical Society, 1981.

Artley, Bob. *A Book of Chores, As Remembered By A Former Kid.* Ames, IA: Iowa State University Press, 1989.

Atkins, Annette. *Harvest of Grief: Grasshopper Plagues and Public Assistance in Minnesota, 1873-1878.* St. Paul: Minnesota Historical Society Press, 1984.

Bidwell, Percy Wells, and John I. Falconer. *History of Agriculture in the Northern United States: 1620–1860.* New York: Peter Smith, 1941.

Blecker, Paulin. *Deep Roots: One Hundred Years of Catholic Life in Meire Grove.* St. Cloud: Sentinel Publishing, 1958.

Brinkman, Marilyn. *Bringing Home The Cows: Family Dairy Farming in Stearns County, 1853–1986.* St. Cloud: Stearns County Historical Society, 1988.

Bryant, Charles S. and Abel B. Murch. *A History of the Great Massacre By the Sioux Indians in Minnesota.* Cincinnati: Rickey and Carroll, 1864.

Buck, Solon J., and Elizabeth Hawthorn Buck. *Stories of Early Minnesota.* New York: Macmillan Company, 1927.

Carley, Kenneth. *The Sioux Uprising of 1862.* St. Paul: Minnesota Historical Society Press, 1976.

Conzen, Kathleen Neils. *Making Their Own America: Assimilation Theory and the German Peasant Pioneer.* New York: Berg, 1990.

Danbom, David B. *Born in the Country: A History of Rural America.* Baltimore: Johns Hopkins University Press, 1995.

Dickerman, Charles W. *How to Make the Farm Pay.* Philadelphia: Zeigler, McCurdy, c. 1869.

Drache, Hiram M. *The Challenge of the Prairie.* Fargo: North Dakota Institute for Regional Studies, 1970.

Fergus Falls, 1872–1972. Fergus Falls: Centennial Corporation, 1972.

Fessenden, Thomas G. *The Complete Farmer and Rural Economist.* New York: A.O. Moore Agricultural Book Publishing, 1858.

Fite, Gilbert C. *American Farmers: The New Minority.* Bloomington: Indiana University Press, 1981.

Franklin Steele, Jr., & Company's Illustrated Catalog of Agricultural Implements. Minneapolis: Johnson, Smith & Harrison Printers, [probably 1881, 1882, or 1883].

Frost, Robert. *Complete Poems of Robert Frost, 1949.* New York: Henry Holt and Company, 1949.

Garfield Centennial Committee. *Garfield: The First 100 Years, 1880–1980*. Fertile, MN: Garfield Centennial Committee, 1980.

Giencke, Rosanne Begantine. Phelps: *A Peek Into Its Past*. Fergus Falls, MN: Rosanne Giencke, 1991.

Guither, Harold D. and Harold G. Halcrow. *The American Farm Crisis: An Annotated Bibliography with Analytical Introductions*. Ann Arbor, MI: Pierian Press, 1988.

Halsted, Byron D. *Barns, Sheds and Outbuildings*. Brattleboro, VT: Stephen Greene Press, 1977.

Hidy, Ralph W., Muriel E. Hidy, and Roy V. Scott with Don L. Hofsommer. *The Great Northern Railway: A History*. Boston: Harvard Business School Press, 1988.

Holmquist, June Drenning, ed. *They Chose Minnesota: A Survey of The State's Ethnic Groups*. St. Paul: Minnesota Historical Society Press, 1981.

Howe, Charles W. *Forty Wonderful Years: Morgan, Minnesota and Vicinity, 1876–1916*. Morgan: Morgan Messenger, 1916.

Humphries, W.R., and R. B. Gray. *Partial History of Haying Equipment*. Information Series, no. 74. Beltsville, MD: U.S. Department of Agriculture, 1949.

Isern, Thomas D. *Bull Threshers and Bindlestiffs*. Lawrence, KS: University of Kansas Press, 1990.

Jarchow, Merrill E. *Like Father, Like Son: The Gilfillan Story*. St. Paul: Ramsey County Historical Society, 1994.

Jarchow, Merrill E. *The Earth Brought Forth: A History of Minnesota Agriculture to 1885*. St. Paul: Minnesota Historical Society Press, 1949.

Johnson, Paul C. *Farm Animals in the Making of America*. Des Moines, IA: Wallace Homestead Book Company, 1975.

Johnson, Paul C. *Farm Inventions in the Making of America*. Des Moines, IA: Wallace-Homestead Book Company, 1976.

Klinkenborg, Verlyn. *Making Hay*. New York: Random House, 1986.

Knight, Edward H. *Knight's New Mechanical Dictionary*. New York: Houghton, Mifflin & Company, 1884.

Kysor, Kenneth. *Time On My Hands*. Cattaraugus, NY: Kenneth Kysor, 1975.

Loehr, Rodney. *Minnesota Farmers' Diaries*. St. Paul: Minnesota Historical Society Press, 1939.

Lovoll, Odd S. *Norwegians On The Land*. Marshall, MN: Society for the Study of Local And Regional History, Southwest State University, 1992.

Marshall, Josiah T. *The Farmer's and Emigrants Hand-Book*. Hartford, CT: O.D. Case & Company, 1851.

Macmillan, Don, and Russell Jones. *John Deere Tractors and Equipment, 1837–1959*. St. Joseph, MI: American Society of Agricultural Engineers, 1988.

McMurry, Sally. *Transforming Rural Life: Dairying Families and Agricultural Change, 1820–1885*. Baltimore: Johns Hopkins, 1995.

Meyer, Kent. *A Witness of Combines*. Minneapolis: University of Minnesota Press, 1998.

Miller, Patricia Marthaler. *Marthaler Genealogy*. Sauk Centre, MN: Self-published, 1979.

Mihelich, Josephine. *Andrew Peterson and the Scandia Story*. Minneapolis: Ford Johnson Graphics and the author, 1984.

Minnesota Historical Society. *Minnesota Farmscape: Looking at Change*. St. Paul: Minnesota Historical Society Press, 1980.

Moberg, Vilhelm. *The Last Letter Home.* St. Paul: Minnesota Historical Society Press, 1995.

New Idea, Incorporated. *Modern Haying.* Coldwater, Ohio: New Idea, Inc., 1936.

Noble, Allen G. *Wood, Brick, and Stone,* vol. 2. Amherst, MA: University of Massachusetts Press, 1984.

Noble, Allen G. and Hubert G.H. Wilhelm, eds. *Barns of the Midwest.* Athens, OH: Ohio University, 1995.

Oxford English Dictionary, 1989 ed. S.v. "Hay."

Oxford English Dictionary, 1933 ed. S.v. "Swath," "Rake."

Parkerson, Donald H. *The Agricultural Transition in New York State.* Ames: Iowa State, 1995.

Peterson, Fred W. *Building Community, Keeping the Faith: German Catholic Vernacular Architecture in a Rural Minnesota Parish.* St. Paul: Minnesota Historical Society Press, 1998.

Rawson, Richard. *Old Barn Plans.* Toronto: Beaverbooks, 1979.

Rider, Allen R. and Stephen D. Barr. *Hay and Forage Harvesting.* Moline, IL: Deere & Company, 1976.

Robinson, Solon. *Facts for Farmers.* New York: A.J. Johnson, 1866.

St. Cloud City Directory, 1910. St. Cloud: John H. Ley, 1910.

Schlebecker, John T. *Whereby We Thrive: A History of American Farming, 1607–1972.* Ames, Iowa: Iowa State University Press, 1975.

Sears, Roebuck and Company. *The 1902 Edition of the Sears Roebuck Catalogue.* Chicago: Sears, Roebuck, 1902; reprint, New York: Bounty Books, 1969.

Sharrow, Gregory, ed. *Families On the Land: Profiles of Vermont Farm Families.* Middlebury, VT: Vermont Folklife Center, 1995.

Simkin, Tom and Richard S. Fiske. *Krakatau 1883: The Volcanic Eruption and Its Effects.* Washington, D.C.: Smithsonian Institution Press, 1983.

Skardal, Dorothy Burton. *The Divided Heart: Scandinavian Immigrant Experience through Literary Sources.* Lincoln: University of Nebraska Press, 1974.

Smith, Harris P. *Farm Machinery and Equipment.* New York: McGraw-Hill Book Company, 1948.

Stearns County Directory, 1910. St. Cloud: Birney Moore, 1910.

Stephens, Henry. *The Book of the Farm,* Volume I. New York: C.M. Saxton, 1855.

Stokes, Samuel N. *Saving America's Countryside: A Guide to Rural Conservation.* Baltimore: Johns Hopkins University Press, 1989.

Tresemer, David. *The Scythe Book.* Brattleboro, VT: By Hand & Foot, LTD., 1981.

Webb, Wayne E. and J.I. Swedberg. *Redwood: The Story of a County.* St. Paul: North Central Publishing, 1964.

Yale, Allen R., Jr. *While The Sun Shines: Making Hay in Vermont 1789–1990.* Montpelier, Vermont: Vermont Historical Society, 1991.

Zetetic Club. *Morgan's Roots Reach 100 Years.* New Ulm: MMI Graphics, 1978.

Magazines

Borcherding, James R. "How to Make Perfect Haylage." *Successful Farming* (June/July 1983): D7.

Borcherding, James and Ron Lutz. "Sealed Storage: Will it Fit Your Livestock System?" *Successful Farming* 69 (June/July 1971): B1–B5.

Brick, Pam. "Fond Memories of the Odor of Plastic." *Omni* 16 (January 1994): 33.

Burke, Jeffrey. "Country Wisdom." *Harper's Magazine* 261 (July 1980): 80–81.

Dyer, Walter A. "Chronicles of a Countryman." *Country Life* 52 (July 1927): 70–76.

The Farmer, 3 July 1943.

Farrell, J.J. "The Growth of Minnesota's Dairy Industry." *Western Magazine* 7 (January 1916): 120

Guebert, Alan. "Randolph County, Illinois: Harder Work, Less Cash, More Debt." *Successful Farming* (January 1984): 19–21.

Magnuson, Ed. "Real Trouble on the Farm." *Time* 125 (18 February 1985): 24–31.

McMillen, Wheeler. "The Ancient Technology of Farming: Ohio, 1910." *American Heritage of Invention & Technology* 7 (Spring/Summer 1991): 44–49.

Mueller, William. "Growing With The Times." *Harper's Magazine* 261 (July 1980): 82–84.

Powell, E.P. "In The Hayfield." *The Outing Magazine* 52 (9 June 1908): 346–349.

"Problems in Agricultural Engineering." *The Farmer* 33 (27 March 1915): 529.

Reed, J.D. "On the Farm: Barn Again!" *Time* (20 February 1989): 87–88.

Sheets, Kenneth R. "America's Farmers Down the Tubes?" *U.S. News & World Report* 98 (4 February 1985): 47–49.

"Sacred Cow." *Time* 125 (18 February 1985): 26.

"Spontaneous Combustion of Hay." *Scientific American* 145 (October 1931): p. 278–279.

"Summer Drinks." *The Farmer*, July 15, 1905: 471.

Tevis, Cheryl. "Farming, America's Most Hazardous Occupation." *Successful Farming* 83 (April 1985): 16.

"The Burning Question." *The Farmer* 23 (August 1, 1905): 495.

Werner, Carl. "Getting in the Hay." *Everybody's Magazine* 29 (July 1913): "What Does Your Memory Smell Like?" *USA Today* 120, (January 1992): 5.

Wilmore, Rex. "The Boom In Sealed Silos." *Farm Journal* 91 (September 1967): 28–29, 49.

Journals

Balmer, Frank E. "The Farmer and Minnesota History." *Minnesota History* 7 (September 1926): 199–217.

Bosch, Roland. "Old Barns Bring Memories (And Pangs)." *Kandi Express, County News [Kandiyohi County Historical Society, Willmar, MN* (October 1993): 9.

Christensen, Thomas P. "Danish Settlement in Minnesota." *Minnesota History* 8 (December 1927): 363–385.

Critchfield, Richard. "Preserving Rural Cultures in the Twenty-First Century." *North Dakota History* 62 (Summer 1995): 34–39.

Drache, Hiram M. "Midwest Agriculture: Changing With Technology." *Agricultural History* 50 (January 1976): 290–302.

Edwards, Everett E. "T. L. Haecker, The Father of Dairying In Minnesota." *Minnesota History* 19 (June 1938): 148–161.

Edwards, Everett E. and Horace H. Russell. "Wendelin Grimm and Alfalfa." *Minnesota History* 19 (March 1938): 21–33.

Fite, Gilbert C. "The Pioneer Farmer: A View Over Three Centuries." *Agricultural History* 50 (January 1976): 275–289.

Gross, Stephen John. "Handing Down the Farm: Values, Strategies, and Outcomes in Inheritance Practices Among Rural German Americans."

Journal of Family History 21 (April 1996): 192–217.

Hendry, George W. "Alfalfa In History." *Journal of the American Society of Agronomy* 15 (May 1923): 171–176.

Jarchow, Merrill E. "Farm Machinery of the 1860's in Minnesota." *Minnesota History* 24 (December 1943): 287–306.

Jarchow, Merrill E. "The Beginnings of Minnesota Dairying," *Minnesota History* 27 (June 1946): 107–121.

Johnson, Hildegard Binder. "Factors Influencing the Distribution of the German Pioneer Population in Minnesota." *Agricultural History* 19 (January 1945): 39–57.

McKnight, Roger. "Andrew Peterson's Journals: An Analysis." *The Swedish Pioneer* 28 (July 1977): 153–172.

Murray, Stanley N. "Railroads and the Agricultural Development of the Red River Valley of the North, 1870–1890." *Agricultural History* 31 (October 1957): 57–66.

Nute, Grace Lee. "The Diaries of a Swedish-American Farmer, Andrew Peterson." *Yearbook of the American Swedish Institute* 1 (1945): 105–132.

Qualey, Carlton C. "Diary of a Swedish Immigrant Horticulturist." *Minnesota History* 43 (Summer 1972): 63–70.

Newspapers

Jameson, Marion K. "Wright Rambles Then and Now." *Howard Lake Herald*, 27 March 1980, in Wright County Historical Society "Agriculture" file, Buffalo, Minnesota.

Battle Lake Review, 28 July 1887.

Bemidji Sentinel, 22 July 1921.

Bemidji Daily Pioneer, 5 October 1921.

Fergus Falls Weekly Journal, January–December, 1883. 4 January 1883, 18 January 1883, 25 January 1883, 15 March 1883, 24 May 1883, 5 July 1883, 12 July 1883; 19 July 1883, 26 July 1883, 2 August 1883, 9 August 1883, 16 August 1883, 23 August 1883, 30 August 1883, 6 September 1883, 20 September 1883, 4 October 1883, 22 November 1883; 27 December 1883. 8 December 1910.

"State of Maine Natives Founded Township in 1871." *Fergus Falls Daily Journal*, 25 July 1968, 1.

"Early Reminiscences of Otter Tail County." *Fergus Falls Journal*, 12 July 1929, 2.

Fertile Journal, December 1956–October 1959. 29 April 1959, 29 July 1959.

"Imdieke Farm Has a History of Making History." *Sauk Centre Herald*, 21 November 1979, B7.

Melrose Beacon, January–December, 1924.

Brown County Journal [New Ulm, MN], 12 August 1927, 18 November 1927, 20 July 1928.

"Minnesota 13–It Turned Corn Into Gold." *St. Paul Dispatch*, 12 October 1962, p. 1, Section 2, in "Prohibition" file, Stearns County Historical Society, St. Cloud, MN.

"Notice," *Minnesota Pioneer*, May 23, 1850, 3.

"Tour Will Take Children to 1862 Conflict." *Star-Tribune*, 9 August 1987, B13.

Minneapolis Star-Tribune. 20 February 1997.

"Effects of Lightning, Barns Burned." *The Monthly Genesee Farmer*, vol. III, no. 9, (September 1838): 137.

Morgan Messenger, January–December, 1968. 7 September 1978, 22 June 1978, 5 September 1979.

New York Times, 13 August 1867, 31 October 1872, 12 November 1877.

"Alfalfa Needs Growth In Fall." *Redwood Gazette*, 17 August 1967, 4.

Redwood Gazette, 1915–1916, 1946–

1947, 1967–1968. 13 June 1935, 22 August 1967.

St. Cloud Daily Times, 26 July 1899, 27 July 1899, 17 July 1931, 21 June 1943, 29 March 1996.

Haga, Chuck. "Science knows the nose has a powerful memory." *Minneapolis Star-Tribune*, 23 July 1992, A1.

"History Loud and Lively at Fort Snelling." *St. Paul Pioneer Press*, 14 June 1999, B4.

"Milk Production Up, Market Share Down." *St. Paul Pioneer Press*, 21 November 1999, A21.

Waconia Patriot, 16 March 1922.

Wheelock's Weekly [Fergus Falls, MN], 8 August 1912.

The Yellowstone Journal [Miles City, MT], 22 December 1883.

Interviews

Berg, Jeff. Interview by author, 7 September 1999, Barnesville, MN.

Dahmes, LaVerne. Interview by author, 27 December 1994, Redwood Falls, MN.

Dahmes, Michael. Interview by author, 7 June 1996, Redwood Falls, MN.

Dickinson, Agnes (Jacobson). Interview by author, 21 July 1996, Bemidji, MN.

Dickinson, Earle. Interview by author, 21 July 1996, 19 March 1997, Bemidji, MN.

Eckblad, Herbert, Vasa Community, Goodhue County, MN. Interview by Ken Peterson, 11 July 1975, Interview #38, transcript. Goodhue County Historical Society, Red Wing, MN.

Engel, Tom. Interview by author, 18 January 1998, Morgan, MN.

Feddema, Herbert, St. Cloud, MN. Interview by Mildred Dumonceaux, 30 January 1979. Interview #1354, transcript. Stearns County Historical Society, St. Cloud, MN.

Felty, Mary Kay (Weber) Felty. Interview by author, 27 July 1996, Morgan, MN.

Hocking, Wendy (Sandmann). Interview by author, 22–23 December 1996, 18 January 1998, South St. Paul, MN.

Hoffbeck, Alvina. Interview by author, 2 June 1996, 17 October 1996, Morgan, MN.

Hoffbeck, Chris George. Interview by author, 22 December 1996, 1 February 1998, Grand Rapids, MI.

Hoffbeck, Dana J. Interview by author, 18 September 1996, 8 February 1998, Zumbrota, MN.

Hoffbeck, Earl. Interview by author, 27 December 1994, Redwood Falls, MN.

Hoffbeck, Francis, and Evelyn (Spong) Hoffbeck. Interview by author, 4 June 1996, Marshall, MN.

Hoffbeck, Jeffrey. Interview by author, 23 December 1996, Morgan, MN.

Hoffbeck, Wayne. Interview by author, 18 June 1996, Woodbury, MN.

Jacobson, Alvin. Interview by author, 29 July 1997, Deer Lake, MN.

Kysor, Kenneth. Interview by author, 5 June 1997, Cattaraugus, New York.

Lee, Arthur O. Interview by author, 24 June 1996, Bemidji, MN.

Lee, Marlys (Rongen). Interview by author, 5 October 1996, Mayville, ND.

Loehr, Edwin, Spring Hill, MN. Interview by Rosie Olmschenk, 9 August 1978. Interview #1105, transcript. Stearns County Historical Society, St. Cloud, MN.

Marthaler, Gary. Interview by author, 13 August 1997, Meire Grove, MN.

Marthaler, Gilbert. Interview by author, 22 July 1996, 11 June 1997, Meire Grove, MN.

Marthaler, Gilbert, Meire Grove, MN. Interview by John Decker, 25 January 1977. Interview #1349, transcript. Stearns County Historical Society, St. Cloud, MN.

Meyer, Elizabeth A. (Nietfield), Melrose, MN. Interview by Mildred Dumonceaux, 29 November 1979, Interview #1436, transcript. Stearns County Historical Society, St. Cloud, MN.

Michalek, Russell. Interview by author, 27 February 1993, Blackduck, MN.

Michalek, Ruth (Needham). Interview by author, 27 July 1997, Blackduck, MN.

Moe, Carlton. Interview by author, 15 November 1998, Barnesville, MN.

Olson, Ron. Interview by author, 29 July 1997, Georgetown, MN.

Palkki, Isaac. Interview by author, 21 February 1998, Grand Rapids, MN

Rongen, Douglas. Interview by author, 25 July 1996, 14 August 1997, Fertile, MN.

Schneider, Michael C., St. Cloud, MN. Interview by Mildred M. Dumonceaux, 25 January 1979, Interview # 1352, transcript. Stearns County Historical Society, St. Cloud, MN.

Sollum, Almond (Al). Interview by author, 21 July 1996, Bemidji, MN.

Stangeland, Jeff. Interview by author, 21 November 1999, Barnesville, MN.

Sweiters, Cyril, New Munich, MN. Interview by Marilyn Brinkman, 6 March 1981, Interview #1675, transcript. Stearns County Historical Society, St. Cloud, MN.

Stern, Anton. Interview by author, 22 July 1996, Belgrade, MN.

Winkels, Charles. Interview by author, 5 December 1999, Rothsay, MN.

Diaries

Allen W. Dawley Diary, 1879, Minnesota Historical Society Collections, St. Paul, MN.

Francis M. Dyer Diary, 1867–1869, 1879, Minnesota Historical Society Collections, St. Paul, MN.

Andrew Peterson Diary Translation in Andrew Peterson Papers, 1854–1961, Minnesota Historical Society Collections, St. Paul, MN.

Oliver Perry Kysor Diary, Maine Township, Minnesota, 1883, Otter Tail County Historical Society Collections, Fergus Falls, MN.

Government Documents

Bureau of the Census. *U.S. Census, 1870.* New York, Cattaraugus County, Leon Township.

Bureau of the Census. *U.S. Census, 1880.* New York, Cattaraugus County, Leon Township.

Bureau of the Census. *United States Census of Agriculture: 1925.* Washington, DC: GPO, 1927.

Bureau of the Census. *Fifteenth Census of the United States: 1930.* Vol. 1, Population. Washington, D.C.: GPO, 1931.

Bureau of the Census. *1992 Census of Agriculture.* Volume 2, Part 3, "Ranking of States and Counties." Washington, D.C.: GPO, 1994.

"Cemetery Records, Otter Tail County, MN, Maine Township, Silent Vale Cemetery," vol. 5. Otter Tail County Historical Society, Fergus Falls, MN, roll 36, microfilm.

1885 Minnesota State Census, Maine Township, Otter Tail County.

United States Department of Agriculture. *Grass: The Yearbook of Agriculture, 1948.* Washington, D.C.: G.P.O., 1948.

United States Department of Agriculture. *Key To The Native Perennial Grasses—Midwest Region East of the Great Plains.* Washington, D.C.: G.P.O., 1968.

United States Department of Agriculture. *Making and Feeding Hay-Crop*

Silage. Washington, D.C.: G.P.O., 1962.

"Feed Storage." University of Alberta website, 1998, http://www.afns.ualberta.ca.

Other Unpublished Materials

Babcock, Charles D. "Kysor Genealogy Compiled by Charles D. Babcock," manuscript, 1968.

Buckingham, Bill. "The Agricultural History of Otter Tail County," in "Agriculture" file, Otter Tail County Historical Society, Fergus Falls, MN.

Department of the Interior. "Andrew Peterson Farmstead." National Register of Historic Places nomination, Minnesota Historical Society, St. Paul, MN, 1978.

Hoffbeck, John L. "Haymaking as Part of My Farming Experience," typescript, 20 August 1995, Cincinnati, OH.

Hudson, Lew. "Harvest Time," Agriculture file, Wright County Historical Society, Buffalo, MN.

Kysor, Kenneth. "The Kysors of Cattaraugus County, New York." March, 1984.

Lee, Marlys (Rongen). "Arthur Rongen's Life," typescript, 1995, Mayville, ND.

Lee, Marlys (Rongen) Lee. "Childhood Memories of Charles for his Children by their Rongen Uncles and Aunts," 1990, typescript, Mayville, ND.

Rogers, Gilbert. "Reminiscences," Goodhue County Historical Society, Red Wing, MN, July 1, 1978.

Setterberg, Lois. "Carolina Svensdotter Johnson, 1828–1901," family history manuscript, n.d.

"Stearns County Agricultural Statistics." "Statistics" file, Stearns County Historical Society, St. Cloud, MN.

Letter. Russ Bjorhus, Litchfield, MN, to the author, 31 October 1996.

Letter. Harlan Dalager, Minnetonka, MN, to the author, 18 December 1996.

Letter. Katherine Gausman, Morris, MN, to the author, 24 October 1996.

Letters. Alvina Hoffbeck, Morgan, MN, to the author, 27 June 1974, 10 June 1975, 13 July 1976, 20 July 1976, 27 October 1976, 26 February 1987.

Letter. Chris George Hoffbeck, Grand Rapids, Michigan, to the author, 2 February 1998.

Letter. Wendy Sandmann Hoffbeck, Morgan, MN, to the author, 2 August 1977, February 1984.

Letters. Gertrude Johnson, Hammond, Wisconsin, to the author, 7 November 1996, 12 December 1996.

Letter. Warren Lindberg, Lake Hubert, MN, to the author, 23 November 1996.

Letter. Gilbert Marthaler, Meire Grove, MN, to the author, 15 August 1996.

Letter. Edith (Wickander) McMillan, Wheaton, MN, to the author, 2 November 1996.

Letter. Modesta "Maudie" (Mueller) Pemberton, Blue River, OR, to the author, 31 December 1996.

Letter. Marlys (Rongen) Lee, Mayville, ND, to the author, 24 March 1997.

Letter. Douglas Rongen, Fertile, MN, to the author, 15 April 1997.

Letters. Gertrude (Baden) Simer, Minneapolis, MN, to the author, 11 November 1996, 5 December 1996.

Letter. John Schillberg, Osceola, WI, 5 December 1996.

Letter. Vernon Shoquist, Chisago City, MN, to the author, 1996.

Letter. Laurance Stadther, Olivia, MN, to the author, 23 November 1996.

Letter. Ernest Swanson, Goodridge, MN, to the author, 29 October 1996.

Letter. Woods, Evalyn (Bornhoft), Sonora, CA, to the author, 20 November 1996.

Theses, Dissertations.

Kline, Gerald L. "Harvesting Hay With The Automatic Field Baler." M.S. thesis, Iowa State University, 1946.

Loomis, Ormond H. "Tradition and the Individual Farmer: A Study of Folk Agricultural Practices in Southern Central Indiana." Ph.D. diss., Indiana University, 1980.

Rosycki, Anthony. "The Evolution of the Hamlets of Stearns County, Minnesota." M.A. thesis, University of Minnesota, 1976.

Index

Frontispiece, 20, 21, 26 (Harry D. Ayer, photographer), 28, 55, 60, 67, 73, 78, 86, 110, 125—Minnesota Historical Society

Page 3, 6, 8, 13, 15, 45, 47, 74, 77, 105, 106, 135, 136, 141, 143, 150, 152, 165, 177—Author's collection

Page 19, 139—Drawing by Matt Kania

Page 30—Beltrami County Historical Society

Page 33, 89, 109, 126, 127—Drawing by Bill Mohn

Page 41—Wabasha County Historical Society

Page 49—Rueben L. Ruth Agricultural Scrapbook #7, Polk County Historical Society

Page 50—Otter Tail County Historical Society

Page 51—Courtesy of the Kysor family

Page 62—Courtesy of Agnes (Jacobson) Dickinson

Page 79, 82, 91, 103—Courtesy of the Marthaler family

Page 87—Courtesy of AGCO Corporation

Page 111, 113, 115, 118, 120— Courtesy of the Rongen family

Page 121—Myron Hall Collection, Stearns County Historical Society Museum, St. Cloud, MN

Page 140—Courtesy of Rockford Map Publishers

The Haymakers was designed and set in type by Will Powers at the Minnesota Historical Society Press. The types are Bell, cut in 1788 by Richard Austin, and Franklin Gothic, designed in 1904 by Morris Fuller Benton. *The Haymakers* was printed by Thomson-Shore, Inc., Ann Arbor.